The Oceanographer's Companion

The Oceanographer's Companion

Essential Nautical Skills for Seagoing Scientists and Engineers

By George A. Maul

CRC Press
Taylor & Francis Group
Boca Raton London New York

CRC Press is an imprint of the
Taylor & Francis Group, an **informa** business

CRC Press
Taylor & Francis Group
6000 Broken Sound Parkway NW, Suite 300
Boca Raton, FL 33487-2742

© 2017 by Taylor & Francis Group, LLC
CRC Press is an imprint of Taylor & Francis Group, an Informa business

No claim to original U.S. Government works

Printed on acid-free paper

International Standard Book Number-13: 978-1-4987-7306-5 (Paperback)

Library of Congress Cataloging-in-Publication Data

Names: Maul, George A., author.
Title: The oceanographer's companion : essential nautical skills for seagoing scientists and
    engineers / George A. Maul.
Description: Boca Raton, Fla. : CRC Press, 2017. | Includes index.
Identifiers| LCCN 2016035580| ISBN 9781498773065 (hardback : alk. paper) |
    ISBN 9781498773096 (ebook)
Subjects: LCSH: Oceanography—Handbooks, manuals, etc. | Seamanship—Handbooks,
    manuals, etc.
Classification: LCC GC24 .M38 2017 | DDC 623.88—dc23
LC record available at https://lccn.loc.gov/2016035580

Visit the Taylor & Francis Web site at
http://www.taylorandfrancis.com

and the CRC Press Web site at
http://www.crcpress.com

Contents

Preface

When you embark on a research vessel, you enter a world very different from the shore-side laboratory or classroom. Many of the words commonly used ashore are not used on-board, and "old salts" will judge your "savvy" accordingly. Learning the language of going to sea is not only important for proper communication, but also for your safety and well-being. This handbook is written to acquaint the landlubber with ships, their traditions, construction, and the work done for you by the officers and crew. It is not meant as a treatise in seamanship, but rather to prepare you for a successful and safe research cruise. It is a "handbook," so toss it into your seabag along with foul-weather gear, closed-toed shoes, sunscreen, work clothes, and reading materials. Be observant! Why did the captain or chief marine engineer do this or that? What do those flags mean? Where is your fire and emergency muster station? Lifeboat station? Can you tie a few essential knots, such as a bowline? How do you pronounce "bowline?" Writing this handbook is the output of teaching oceanographers, ocean engineers, marine meteorologists, and other students at Florida Institute of Technology, and it is for them and their fellow future shipmates from all marine science institutes that this is intended. Bon Voyage!

George A. Maul
Melbourne, Florida

Acknowledgments

Writing a book is a team effort that ranges across many disciplines and areas of expertise, some of which goes back to my undergraduate years in the New York Maritime College at Fort Schuyler, others to years as a ship's officer in the US Coast and Geodetic Survey, still others as a seagoing scientist on a variety of research vessels. Almost all of the illustrations are the work of Robert Gribbroek in the Florida Institute of Technology's Creative Services office; his skill, artistry, and contributions are greatly appreciated. William Battin, the extraordinary field engineer of Florida Tech's Department of Marine and Environmental Systems, supported much of the photography of Dominic Agostini and cared for many of the oceanographic instruments of historical value. My editor, Irma Britton, contributed to the final product, and in fact suggested the title. My wife Carole truly is "the oceanographer's companion" and has been for over 50 years. My ultimate inspiration comes from the psalmist who in Psalm 107 (KJV) writes:

23 They that go down to the sea in ships, that do business in great waters;

24 These see the works of the LORD, and his wonders in the deep.

25 For he commandeth, and raiseth the stormy wind, which lifteth up the waves thereof.

26 They mount up to the heaven, they go down again to the depths: their soul is melted because of trouble.

27 They reel to and fro, and stagger like a drunken man, and are at their wit's end.

28 Then they cry unto the LORD in their trouble, and he bringeth them out of their distresses.

29 He maketh the storm a calm, so that the waves thereof are still.

30 Then are they glad because they be quiet; so he bringeth them unto their desired haven.

31 Oh that men would praise the LORD for his goodness, and for his wonderful works to the children of men!

Author

George A. Maul earned his BS (with honors) from the New York Maritime College, Fort Schuyler, and was licensed a US merchant marine officer; he received his PhD from the University of Miami, where he later taught as adjunct professor of meteorology and physical oceanography. LCDR Maul served 9 years as a commissioned officer in the US Coast and Geodetic Survey and became a *Shellback* while serving as operations officer aboard the USC&GS Ship *Discoverer*. He then worked for 25 years as a research oceanographer and supervisory oceanographer with the National Oceanic and Atmospheric Administration, receiving three distinguished authorship awards during his NOAA tenure. Dr. Maul currently is professor of oceanography at the Florida Institute of Technology, where for 20 years he was department head of Marine and Environmental Systems; he is a recipient of faculty senate excellence awards for service and for teaching. He has been chief scientist on numerous oceanographic cruises and has published over 200 refereed articles and book chapters on oceanography and meteorology, editorials, technical reports, refereed abstracts, and books. Prof. Maul is a Fellow of the Marine Technology Society, a Fellow of the American Meteorological Society, and the 2016 Medalist of the Florida Academy of Sciences.

1

Research Vessels of the Past and Present—A Brief History of Seagoing Science

Humans are explorers and have ventured far and wide since leaving our primordial home in Africa some 100,000+ years ago. Watercraft, undoubtedly, have been one of the many tools our ancestors used to cross rivers and lakes; crossing oceans took a bit longer for the technology to meet the challenge, but there is some archeological evidence of crossing short ocean distances 130,000 years ago. Rafts and other nondisplacement vessels most likely were the first watercraft, dugouts the first boats (the oldest known dugout, the *Pesse canoe*, is dated *ca.* 8000 BC), followed by reed craft and eventually ships powered by wind and oar and motor. Reed boats are depicted in petroglyphs that could date to 12,000 years ago; the oldest remains of a reed boat is about 7000 years old and was found in modern-day Kuwait. Egyptian drawings of reed boats date to *ca.* 4000 BC, and Thor Heyerdahl theorized that such a craft could have crossed the Atlantic Ocean by sailing *Ra II* from Africa to the Caribbean in 1970. A Roman amphora dating to *ca.* 200 BC was found in Brazil by Sir Robert F. Marx, so pre-Christian knowledge of the Americas is quite possible.

Egyptians, Mesopotamians, Greeks, Romans, Phoenicians, Carthaginians, Japanese, Indians, Arabs, Chinese, Vikings and other Europeans, Polynesians, indigenous Americans, and the list goes on, have explored the oceans. Their ships evolved from reed boats to outriggers to wooden displacement hulls until the mid-1800s when metal began to be used. Many ships of discovery and their names are unknown, but replica vessels have been built and sailed. Reed was used not only to build boats and rafts, but also for cordage and sails, as well as papyrus for writing; reed boats and sails are still found on Lake Titicaca in South America. Mesopotamians are credited with inventing sailboats and probably made the first wooden boats; these were displacement hulls and the wooden planks were sewn together. Greek and Roman shipbuilding certainly advanced the art and forged the ability to sail outside the Mediterranean Sea, but lands and sea routes were actively being discovered and fought over back to at least 1210 BC, according to the saga of the "Sea Peoples."

So are these or any other ships of discovery properly called oceanographic research vessels (RVs)? Little was quantified about Earth before Greek

FIGURE 1.1
Greek geographers of the sixth century BC were using information gathered by seamen to construct the earliest maps of the coastline. This reconstructed map is attributed to Hecataeus of Miletus *ca.* 500 BC, who was a student of Anaximander (*ca.* 610–546 BC) to whom the original map is credited. ΩΚΕΑΝΟΣ is Oceanus, ΕΥΡΩΠΗ is Europa, ΛΙΒΥΗ is Libya (all of Africa was called Libya at the time), and ΑΣΙΑ is Asia.

philosophers took to drawing fairly accurate coastline maps, such as shown in Figure 1.1. Earlier examples of map-making date to the Babylonians; Sargon of Akkad (*ca.* 2300 BC) seems to have predated Greek ideas of a flat Earth surrounded by ΩΚΕΑΝΟΣ (Oceanus) as also proposed by Homer (*ca.* 850 BC). Yet it was Thales of Miletus (*ca.* 585 BC) who understood Earth to be a sphere and invented the gnomonic projection (Figure 7.1) to map a sphere on a flat surface. The projection of Figure 1.1 is unknown, but clearly information gathered by ships and sailors with an understanding of latitude and longitude was a research effort.

Timekeeping

Seagoing scientists and engineers know that determining longitude requires precise timekeeping. John Harrison's marine chronometer, *ca.* 1760, needed a known rate of change, and this was accomplished by comparison with chronometers at local observatories. In 1829, Royal Navy Captain Robert

Wauchope invented the time ball, first deployed in Portsmouth, England. The time ball was made to be dropped, guided along a central pole, in a tower at the observatory. At exactly 1 PM the ball was released to signal Greenwich Mean Time at the instant the ball began to fall. Ships in the harbor could compare their chronometer with the observatory and calculate a correction and rate of correction change. The US Naval Observatory introduced the time ball in 1845, and the idea was in use worldwide until the invention of radio time signals in the 1920s. The time ball in New York City is an outgrowth of this invention.

Oceanography Origins

Oceanography as a word was coined in 1859 by Charles Wyville Thompson and John Murray, both of whom were central to the 1872–1876 expedition of *HMS Challenger*. So was "graphing the ocean" prior to that famous voyage not science? Were the vessels of discovery by Leif Erikson (ship's name unknown), or Sir James Cook (*HM Bark Endeavor*), or Sir Charles Darwin (*HMS Beagle*), or Ferdinand Magellan (*Nao Victoria*), or Jason (*Argo*—the oldest ship name we have), or Fridtjof Nansen on the *Fram*, or US Navy Lieutenant John Elliot Pillsbury aboard the Coast Survey Steamer *Blake*, or … not RVs? Arguably they are, considering how little about the ocean is known even today. For the purposes of this chapter, RVs are defined as ships of the sea (to distinguish them from airships or spacecraft) whose purpose is discovery. In all fairness though, it is the *HMS Challenger* that is credited with machinery and instruments meant to study the underwater realm in particular, and the surface in general, but the first ship specifically built by any government for deep-sea marine research is the *RV Albatross* of the US Commission of Fish and Fisheries in 1882.

In their classic 1942 book *The Oceans: Their Physics, Chemistry, and General Biology*, authors H.U. Sverdrup, Martin W. Johnson, and Richard H. Fleming list four features that are desirable in RVs: (1) sturdiness and seaworthiness, (2) low freeboard, (3) sails, and (4) sufficient deck space. All but one of these features are common in modern science ships, but they all are present in the *RV Albatross* (Figure 1.2). The *Albatross* was steel hulled, had low freeboard (the vertical distance from the main deck to the water), had masts for sails, and being 234 feet long, had all the needed deck space for her work; her beam was 27.5 feet and she displaced 1074 tons. The *RV Albatross* was fitted throughout with electric lights with the generator being designed by none other than Thomas Edison. Being that she operated before the invention of echo sounders for determining water depth, the *Albatross* was fitted with two sounding machines: a Sigsbee Sounding Machine connected to a steam-powered winch capable of hauling in wire at the rate of 100 fathoms

FIGURE 1.2
RV Albatross, first ship designed and built to be a research vessel. (Photo from National Oceanic and Atmospheric Administration library; http://www.nefsc.noaa.gov/history/ships/albatross1/photo3.jpg).

per minute (1 fathom = 6 feet), and a Tanner Sounding Machine for navigational purposes used in water less than 200 fathoms.

The *RV Albatross* had two large laboratories and sufficient storage space for thousands of specimens. She was designed for deep-sea dredging and had 4500 fathoms of 3/8-inch wire rope that ran through a specially designed winch and take-up reel system. She used a Sigsbee water bottle to collect samples for chemical analysis in the onboard laboratory, Negretti and Zambra deep-sea thermometers, and Helgard's ocean salinometer to measure sea water density; these were state-of-the-art instrumentation in the late 1800s. *Albatross* surveyed the sea floor from Newfoundland to the Philippines, Alaska to Cuba. Several of the earlier cruises were under the direction of Professor Alexander Agassiz, and many of the specimens collected went to the Museum of Comparative Zoology at Harvard University. *Albatross* worked until October 1921 when she was decommissioned at Woods Hole, Massachusetts. Truly a remarkable vessel during the transition from sail to steam.

Sverdrup et al., in their chapter on Observations and Collection at Sea in *The Oceans*, had a table summarizing important historical RVs of the era up to *ca.* 1942. Following their lead, Table 1.1 is a modified version of their table from today's perspective. Some historic ships listed in Table 1.1 are duplicated from *The Oceans*; some are not. Many early twentieth century ships in the United States were owned and/or operated by the large oceanographic institutions and universities, whereas outside of the United States most ships were government owned and operated. Expeditions such as the sloop *Vostok* commanded by Fabian Gottlieb Thaddeus von Bellingshausen, and the ship *Williams* under Edward Bransfield, were funded by the governments of Russia and England, respectively. In the United States, most ships were

TABLE 1.1

Sampling of Historic Research Vessels of the Last Few Centuries (Not a Complete List by Any Means)

Name	Nationality	Operator	Hull	Launched	Length	Tonnage	Crew	Scientists
Challenger	United Kingdom	Navy	Wood	1858	225	2137	243	
Albatross	United States	Fisheries Commission	Steel	1882	234	1074		
Fram	Norway	Nansen	Wood	1892	127	402	16	
Meteor	Germany	Navy	Steel	1915	233	1200	114	10
Willebrord Snellius	Netherlands	DoD	Steel	1928	204	1055	84	6
Dana	Denmark	Danish Commission	Steel	1917	138	360	14	8
Kainan Maru	Japan	Private	Wood	1910	100	204	27	
Calypso	France	Cousteau	Wood	1942	139	360	27	

commanded by officers of the Navy, and were mostly converted to oceanographic use.

Ships of the US Coast and Geodetic Survey were primarily hydrographic survey vessels engaged in nautical charting until the commissioning of the USC&GS Ship *Oceanographer* in 1966. USC&GS hydrographic survey vessels were well capable of oceanographic observations, but their primary mission was chart making. The *Oceanographer* was specifically designed as a RV and had more deep-sea capability than any other USC&GS ship; she was at the forefront of a central computer system usable not only for analyzing and storing scientific data, but also for engine room control, navigation, and plotting. *Oceanographer* was commissioned by President Lyndon B. Johnson and sponsored by Lady Bird Johnson at the Washington Naval Yard; Lieutenant Commander George A. Maul was the Officer of the Deck on that day, July 13th. *Oceanographer* served as the queen of the federal research fleet for over 30 years, made an around-the-world voyage as her first significant research cruise, and had a sister ship, the *USC&GSS Discoverer*; both are now retired and decommissioned. USC&GS ships are presently all part of the National Oceanic and Atmospheric Administration (NOAA) fleet.

The US Navy, and the navies of many other countries, also have fleets of survey vessels, many of which can be classified as RVs. Their purposes are parallel to those of the US civilian fleet, but often their mission is classified. The US Naval Oceanographic Office (NAVOCEANO) operates six 100-m-length ships named: *Pathfinder, Bowditch, Heezen, Sears, Sumner,* and *Henson.* Several of these names should be familiar to oceanographers and ocean engineers; they are all T-AGS 60 Class ships, capable of many research missions including bathymetric surveys; T-AGS is a navy designation for "tactical–surveying ship." These ships have the unique mission of having no home port and operate 365 days a year every year. NAVOCEANO personnel are rotated so that the ships operate as oceanographic survey vessels continuously. If a naval ship is designated as USS, it means US Ship and is primarily a warship. If designated as USNS, it means US Naval Ship and is a support vessel not designed for combat. The six NAVOCEANO ships are all USNS, and are operated by civilian crews of the Military Sealift Command.

United States Academic Research Vessels

Most large non-NOAA oceanographic vessels in the United States are scheduled by the University-National Oceanographic Laboratory System (UNOLS), chartered in 1972. UNOLS vessels are typically owned by the US Navy or the National Science Foundation, and are classified as: global class ships (> 235 feet in length), ocean/intermediate class ships (150–235 feet), regional class ships (126–150 feet), and coastal/local class (< 126 feet). Also scheduled in cooperation with UNOLS is the NOAA Ship *Ronald H. Brown* (global class ship) and US Coast Guard icebreakers, all about 400 feet in length. UNOLS ships are designated for federally funded projects. UNOLS has 17 RVs in its fleet, operating from 14 institutions; there are 62 member institutions in the System.

RVs such as the *Oceanographer* and the *Brown* are specialized passenger vessels equipped to conduct a wide variety of oceanic and marine atmospheric experiments. They typically are twin propeller ships with bow and/or stern thrusters, numerous laboratories, computer rooms, refrigerators and freezers, and specialized deck machinery for handling buoys and other heavy instruments such as dredges, nets, corers, submersibles, conductivity-temperature-depth (CTD) systems, and so on. In addition, there are several highly specialized ships such as the *GSF Explorer*, a deep-sea drilling vessel owned by the GlobalSantaFe Corporation, *Flip*, a 355-foot-long semisubmersible research platform owned by the Office of Naval Research and operated by the Scripps Institution of Oceanography, and the *NR-1*, a deep submergence vessel operated by the US Navy and the first nuclear research dive vessel in the American fleet (its depth capability is still classified but most likely not capable of a Mariana Trench dive of 36,000 feet credited to the bathyscaphe *Trieste*; Figure 1.3).

While vessels as described earlier will continue to be part of future fleets, regardless of the country under which they are flagged, many more ships of the future will most likely be autonomous vehicles. Oceanographers have had many such "vessels" in the form of satellite-tracked buoys using ARGOS and other geolocation technology. Added to ARGOS-tracked Lagrangian

FIGURE 1.3
Bathyscaphe *Trieste* was 60 feet in length, weighed 50 tons, had a two-man crew in the pressure sphere at the bottom, and was filled with gasoline for buoyancy. She was retired from service in 1966 and is on permanent exhibit at the National Museum of the US Navy, Washington DC. (Photograph courtesy of the U.S. Naval Undersea Museum).

drifters are more complicated systems such as gliders (underwater robots capable of diving, surfacing, and measuring variables such as temperature, salinity, pressure, etc.). Advanced buoys too can be considered RVs as they are capable of navigation, measuring, and real-time reporting of critical information for forecasting ocean variables and hurricanes, tsunamis, and rogue waves. Buoys, gliders, autonomous surface vehicles, and futuristic submersibles that can fly all ultimately will need servicing and recovery; that will continue to be a major requirement which will keep classic surface vessels in the fleets.

A fellow research scientist once quipped that "oceanography is a rich country's hobby." Most certainly he was thinking (in part) of the cost of running RVs. Hundreds of dollars a day for a small (< 30 feet) inshore vessel to tens (many tens) of thousands of $US a day for a global class ship. Little wonder that most of the work is supported by governments, and yet, given the central role of Earth's seas to humanity—food, pleasure, transportation, security, inquiry—should we not expand the number and role of RVs in our oceans? When, not if, the next *Deep Water Horizon* oil spill event (2010) occurs, only RVs—ready, willing, and able—will provide our computers with the data needed to forecast the fate of flotsam and jetsam. Human health, and that of our coinhibitors on this planet, depends on a fleet of RVs prepared to provide critical data and scientific observation, in a timely and unbiased manner.

Additional Reading

Heyerdahl, T. 1979. *Early Man and the Ocean: A search for the beginning of navigation and seaborne civilizations*. New York: Doubleday & Co, 438 pp.

Exercises

1. Conduct an Internet search for a pre-World War II RV of historic interest (other than those listed in Table 1.1), and (1) write a paragraph about it and (2) make a one-slide PowerPoint that includes a photograph or drawing. List the vessel's dimensions and ownership.
2. Name an oceanographic institution or major university department or school, and (1) write a one-paragraph history of the organization and (2) make a one-slide PowerPoint summary.
3. The following figure shows an item from the eighteenth century used on a ship; it is about 3 feet long. What is it?

2

Seagoing Skills—Crews, Scientific Party, Logistics, Emergencies, First Aid, and Cardiopulmonary Resuscitation

---

The last sail-only oceanographic expedition left Virginia in the summer of 1838. It was called the "Ex Ex" for Exploring Expedition, and was a bold 4-year investment by a young nation to map areas of Antarctica, the Pacific Northwest, and far-flung islands. Lieutenant Charles Wilkes was the commander of the six-vessel fleet; his name is synonymous with numerous geographic sites. The "Ex Ex" circumnavigated Earth and made many scientific observations, over a decade before the term "oceanography" was coined. Wilkes's ships were naval vessels, entirely made of wood, and masterfully sailed using skills that harken back to the Egyptians, Phoenicians, Greeks, Romans, and Vikings. The introduction of steam power and steel in the mid-nineteenth century was a paradigm shift in seamanship.

George L. Riser, a master mariner who made the "Murmansk Run" three times during World War II (and lived to tell about it!), proudly claimed that seamanship is the world's second oldest profession. As in times past, the art of seamanship today includes knowledge of nautical terms, basic ship construction, knots and splices, line handling, navigation, firefighting, damage control, piloting, steering, engines and sails, deck machinery, wires and wire ropes, search and rescue, communications, boat handling, fueling, maneuvering, emergencies, first aid, safety, anchoring, handling of scientific instruments, and the list goes on. On modern research vessels, many historical sea arts are no longer practiced or needed, but the fundamental skills are unchanged, and the seagoing scientist or engineer is well advised to know them.

Mr. Riser (Figure 2.1) was the first lieutenant on a ship operated to train future merchant marine deck and engine officers. "What is a first lieutenant?" is an obvious question for a seagoing scientist or engineer embarking on his or her first voyage. For that matter, what is the organizational structure of a research vessel, and why does it matter? As one might guess, Mr. Riser is in the uniform of the US Maritime Service (USMS). The USMS was instrumental in training US Merchant Marine officers, particularly during World War II, and today is a voluntary organization within the US Maritime Administration. Few, if any, civilian research vessels of the

FIGURE 2.1
Lieutenant George L. Riser; US Maritime Service, master mariner; seaman extraordinaire; patient teacher. (Photo courtesy of the Stephen B. Luce Library and Archives.)

United States will have uniformed USMS officers, that being the niche of several colleges and academies such as the US Merchant Marine Academy or the New York Maritime College at Fort Schulyer. Yet all research vessels will have an organizational structure, and if it is a US Coast Guard (USCG) icebreaker for example, the ship will likely have a first lieutenant.

Shipboard Organization

The position of first lieutenant is shown in Figure 2.2, and is typical of a ship with commissioned officers in command. Commissions are granted to individuals by an act of Congress, signed by the president, and carry naval ranks from ensign (O1), lieutenant junior grade (O2), lieutenant (O3), lieutenant commander (O4), commander (O5), captain (O6), and then to the "flag ranks" of admiral. The commanding officer (CO) can be of any rank, but he or she is usually called "captain" to honor the position. The executive officer (XO) is second in command, and is responsible for most of the day-to-day operations. The scientific party is headed by the chief scientist, and may fit in the organization chart aside, but below the XO. The chief scientist ordinarily will work closely with the operations officer, but always under the command of the CO and XO.

Commissioned ships such as the USCG and National Oceanic and Atmospheric Administration (NOAA) may have a physician and medical staff assigned, especially if there is going to be a long voyage or one to a remort area such as the Arctic Ocean. The sixth and seventh commissioned officer corps of the United States are the NOAA Commissioned Officer Corps and the US Public Health Service Commissioned Corps. These are uniformed officers, but not military officers such as in the air force, army, coast guard, marine corps, and the navy. The ranks are identical to naval ranks (ensign to vice admiral), the pay is the same, the benefits are the same, but they are not

Commissioned research vessel organization

FIGURE 2.2
Generalized organization chart of a ship having commissioned officers in the command. This chart is typical of a US Coast Guard (USGC) Cutter or a National Oceanic and Atmospheric Administration Research Vessel. In addition, there may well be a US Public Health Service medical officer and a USGC weapons officer, and so on depending on the ship's mission.

Civilian research vessel organization

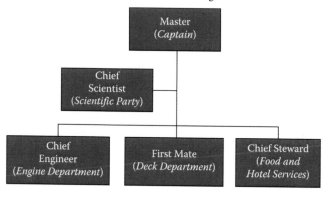

FIGURE 2.3
Organizational structure of a civilian research ship such as those of the University-National Oceanographic Laboratory System (UNOLS). UNOLS ships are owned by the National Science Foundation, and operated by institutions such as the Woods Hole Oceanographic Institution and the Scripps Institution of Oceanography. The medical officer on this class ship is often the first mate, and there may be other professionals depending on the exact ship (dive officer or drilling officer, etc.).

warfare personel, although they may be in harm's way especially during a national emergency such as a general war.

The organizational structure of a civilian research vessel, such as one of the University-National Oceanographic Laboratory System (UNOLS) ships, is somewhat different from the structure of a USCG cutter (USCG ships are called "cutters") or of a NOAA ship. While the details differ, the general issue of line of authority or command is quite clear, and for the very good reason of safety of life at sea (SOLAS). The generalized organization chart of a UNOLS ship is shown in Figure 2.3.

The ship's master is the ultimate authority on a UNOLS ship, and he or she is licensed by the coast guard as a Master of Steam or Motor Vessels of Any Gross Tons, Upon Oceans. US Navy AGORs (Auxiliary General Purpose Oceanographic Research Vessel) are also civilian in organizational structure, and are often classified as global class ships. The *RV Neil Armstrong*, AGOR-27 for example, is 238 feet long, has a beam of 50 feet, and displaces 3043 long tons. She has a crew of 20 and is capapble of carrying 24 scientists, engineers, and technicians in the scientific party.

Crew members on research ships are generally in three specializations: engine, deck, and steward. Under the chief engineer are USCG licenced first, second, and third assistant engineers, and nonlicensed personnel such as electricians, mechanics, oilers, and wipers. The deck department will also have USCG licenced officers (first, second, and third mates), and nonli-censed able-bodied seamen, ordinary seamen, and the senior nonlicensed deck department seaman, the boatswain (pronounced bos'n). The stewards include cooks, bakers and galley workers, accountants, cleaners, and supply personnel.

The person in charge of the bridge on a ship with commissioned officers is called the officer of the deck (OOD). He or she may have a less-experiecned officer assisting who is called the junior officer of the deck (JOOD). The OOD serves in the same capacity as the mate on watch on a UNOLS-structured ship; the mate is licensed by the USCG to serve in this capacity. The OOD is fully trained to the same level as the USCG licensed mate, and shares the same responsibility as authorized by the captain. A similar arrangement exists in the engine room with licensed assistant engineers, either first, second, or third assistant, depending on experience and examination, reporting to the chief engineer.

The Scientific Party

The scientific party is headed by the Chief Scientist. For most work purposes, you will report to him or her, who will assign you to a "watch." A watch is for a specific period of time and specific length of time every day. Many in the ship's crew will be working four on and eight off. This means they typically rotate through the day and night in 4-hour shifts: 0000–0400, 0400–0800, 0800–1200, 1200–1600, 1600–2000, and 2000–2400. So, if someone has the "mid-watch," they will be on duty from 0000 (midnight) to 0400 (4 AM) and again from 1200–1600 (afternoon watch). It is customary to report for duty at least 10 minutes early so as to become familiar with the situation, ask questions, allow time for night vision to form, and allow the watch being relieved to leave on time. This 10 minutes is a time-honored tradition, and not an option! Since the science party is not necessarily "standing watch," that is on their feet for 4 hours straight, the chief scientist may assign different watches such as six on and six off or 12 on and 12 off.

As an aside, the times aforementioned are spoken as follows: 0400 = zero four hundred, 1600 = sixteen hundred—never sixteen hundred *hours* ... and so forth. And when off duty, be respectful of your neighbors' watch and do not disturb his or her sleep.

When preparing to go to sea, all that you need must be taken with you, especially prescription medicines. Most larger ships will have a laundry, so weeks of clothing are not necessary; 5 or 7 days worth will do. Wear only closed-toed shoes and have raingear (never an umbrella) as oceaographic stations are taken 24 hours a day, rain or shine. Remember that sunburn at sea can happen quickly as you will receive both direct sunlight and rays reflected from the sea. A hat and sunscreen are important items in your sea-bag. Some ships will require hardhats and/or steel-toed shoes when on deck and handling equipment.

If a personal conflict arises, as a member of the scientific party, have a private conversation with the chief scientist. Living and working at sea and in such close quarters can be irritating. Most issues can be resolved with an adult-to-adult discussion. Nevertheless, sometimes a "sea lawyer" is among the party (or the crew), and often this mindset leads to unnecessary discontent. Keeping the "esprit de corps" high is everyone's job, but if a legitimate issue does influence harmonious working relationships, dealing with it in a timely manner and with respect to all concerned is a must. UNOLS offers training for seagoing scientists and engineers who aspire to become chief scientist, and all interested should take advantage of this. Most likely, it will be on the job training that develops the next generation of leaders; *leadership by example* is the core method of such professional growth.

Emergencies

As a member of the scientific party, you will be required to particpate in fire and emergencey drills, and abandon ship drills, once a week. Near your bunk (ships do not have beds) there will be an emergency drill "bunk card" (Figure 2.4). On it will be your "muster station" and your lifeboat or life raft station. When you hear a continuous ringing of the ship's emergencey alarm system for at least 10 seconds, you are required to go to your muster station, that is, the location where you will meet and have attendence taken. Usually, the scientific party will all muster in the same location, say the dry laboratory or the mess (ships do not have dining rooms).

The signal for abandon ship is at least six short blasts (1 second each) on the ship's whistle and general alarm followed by a prolonged blast

Event	Fire and emergency	Abandon ship	Man overboard
Signal	▬▬▬▬▬▬ Continuous ringing of alarm bell and a continuous blast on the whistle for 10 seconds	●●●●●●●▬▬ More than six short blasts and one prolonged blast of whistle and alarm bell	▬▬ ▬▬ ▬▬ Three prolonged blasts (4–6 seconds each) of ship's whistle and general alarm bell
Action	Muster in dry lab	Muster at life raft 3	Report to flying bridge
Duties	Assist as directed	Assist as directed	Lookout

FIGURE 2.4
Example of a bunk card that would be found at the foot of your bunk. Having a vertical escape route from your stateroom to the muster station is practicing good seamanship.

(4–6 seconds). On hearing this signal, go directly to your lifeboat or life raft and follow the orders of the crewman in charge. In either case, fire and emergency or abandon ship, always bring your lifevest, wear shoes, and wear a hat. One of the ship's officers will be in charge of the drill or actual emergency, and all members of the sceintific party are to follow his or her orders.

It is essential that you identify your muster station and lifeboat as soon as you embark. Lifeboats are numbered from fore to aft, with even numbers always on the port side, and odd numbers always on the starboard side. Remember that the escape route from your stateroom to your muster station or lifeboat may be vertical. Self-inflating life rafts, such as drawn in Figure 3.4, are designed to launch even if the ship is underwater. They have a pressure release valve integrated into the tie-down straps, which has a manual release button as well. These life rafts are cradled to reach the water if they are released before the ship sinks, even on a listing ship. They are manufactured by Elliot, Viking, Switlik, and others, and are SOLAS (see Chapter 4) and USCG approved.

The signal for man overboard (MOB) is three prolonged blasts on the ship's whistle and on the emergency alarm system. If you spot a person who has fallen overboard, yell "man-overboard starboard" or "man-overboard port" and keep your eyes on him or her. Throw a lifering, and continue to yell and point in the direction of the victim. Do not jump overboard yourself; that will only make two drowning persons. The ship will perform what is known as a "Williamson Turn" (Figure 2.5) to take the stern away from the victim and then reverse course. Continue to point to the victim as the ship eventually turns 180° and returns for the rescue.

Dismissal from drills is three short blasts on the whistle and three short rings on the emergency alarm bells (see bunk card—Figure 2.4).

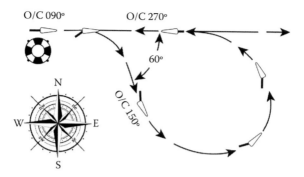

FIGURE 2.5

Williamson turn for man overboard (MOB) on starboard side; lifering shows victim's initial position. Weather and seas permitting, the procedure is to stop all engines immediately, order "right full rudder", come to course XXX° (60° to right of original course)," when 60° to right, order "left full rudder", come to reciprocal course (180° of original course), order "rudder amidships" and "all ahead slow." In the illustration, originally the ship is on course (o/c) 090° (due east). For MOB on port side, the initial order is "left full rudder", and so on. If the ship is steering by magnetic compass, the deviation table will give the correction needed to convert magnetic course to compass course (Chapter 6).

First Aid and Cardiopulmonary Resuscitation

About 12 nm offshore on a research training cruise, it became clear that "Emily" was very seasick, and could hardly stand up. The 85-foot steel-hulled ship had a speed of about 7 knots, and it would take several long hours to get to port. The weather was pleasant, seas were 3–4 feet, light winds, with a gentle ground swell with a period of 10 seconds; the ship was "sea-kindly." Emily could not keep anything down and was pale and nauseous; all feared she might become dehydrated and/or go into shock. We kept her comfortably warm, in the shade, feet elevated, and hydrated, but she was quite ill. The captain called the coast guard on very high frequency (VHF) Channel 16, the emergency hailing frequency, and they dispatched a fast boat to meet us. Emily was transferred to the USCG cutter, taken back to port, hospitalized for the afternoon, and given an intravenous saline solution. She is quite well now, but not going to be a seagoing scientist or engineer.

On a research vessel, if there is no medical officer or corpsman aboard, one of the ship's officers, usually the first mate on a UNOLS vessel, is responsible for first aid and cardiopulmonary resuscitation (CPR). As a member of the scientific party on a larger ship, there may never be an occasion to use medical skills; on a vessel 65 feet or less in length, and especially a runabout, oceanographers and ocean engineers need to be prepared. That means having a first aid kit, basic "first responder" skills, and a plan of action.

A research vessel is designed to be as safe as possible, but going to sea is fraught with some degree of danger. The most likely medical emergencies are choking, bleeding, broken bones, burns, and shock. Before administering first aid, be SAFE—that is, stop, assess, find, and exposure. Stop and think before acting. Assess the situation; do not become the second victim. Find the first aid equipment needed. Exposure: protect yourself—eyes, mouth, nose, hands. Every first aid emergency is different, and as a first responder the victim needs to be handed over to competent medical professionals as quickly as possible. If close to shore, dialing 911 is a critical first step: get help. A quick reference summary is in Table 2.1.

If CPR is needed, such as if a victim has had a heart attack or a near drowning, remember ABCD: airway, breathing, circulation, and defibrillation (Figure 2.6). Lie the victim down, and if conscious ask if they are OK? If unsure, tap the victim on the sternum, and tell them you are a first responder. Check to see if the airway is clear by tilting back the forehead and pressing on the chin to open the mouth. Check to see if they are breathing by placing your ear near their

TABLE 2.1

Common Shipboard Medical Emergencies, Symptoms, and First Aid

Event	Signs	Action
Allergic reaction	Trouble breathing, dizziness, swollen tongue	Assist with victim's EpiPen if needed, look for medical jewelry.
Bleeding	Internal or external	External—apply sterile dressing but no tourniquet unless unable to stop bleeding with compression. Internal—lie victim down, keep victim still, prepare to treat for shock.
Broken bone	Exposed bone or not exposed	Exposed bone—cover with sterile dressing, do not splint. Not exposed—immobilize limb with splint.
Burn	Small or large	Small—cool with cold tap water, no topical ointment. Large—constant cold water, cover with blanket, treat for shock.
Choking	Breathing stopped	Dislodge object by upward thrusts of your fist below sternum from behind.
Electrical burn	Holding object	Do not touch victim, throw circuit breaker or unplug, prepare for cardiopulmonary resuscitation (CPR).
Heart attack	Chest pain, spreading pain to arms, shortness of breath	Chew and swallow an aspirin unless allergic, lie victim down, begin CPR; if automatic external defibrillator (AED) available and patient is unconscious, follow instructions.
Heat exhaustion	Faint, fatigued	Cool victim, remove clothing, water to drink.
Impalement	Penetrated skin	Leave object impaled, do not remove.
Shock	Faint, nauseous, pale or clammy skin	Lie victim down, elevate feet, and keep warm.

FIGURE 2.6

ABCs of cardiopulmonary resuscitation. (a) Check to see that the airway is open and free of obstructions. (b) Check to hear if the victim is breathing. (c) Begin chest compressions at the rate of 100 per minute. Every 30 compressions, give two breaths mouth to mouth, and continue with 30 more compressions. Repeat 30 + 2 cycle until victim revives or help arrives.

nose and mouth and listen for at least 10 seconds. Chest compressions restore circulation by pumping the heart, which may be fibrillating (pulsing rapidly); check for a pulse at the wrist if needed. If an automatic external defibrillator (AED) is on board, ask that it be brought to the victim.

Kneel down with the victim at a right angle to you. CPR chest compression is done by placing the heel of your hand on the victim's sternum and pressing down with both hands, at a rate of 100 compressions per minute. The chest should compress at least 2 inches (5 cm). Use your body weight to press down with elbows locked. After 30 compressions pinch the nostrils, cover the victim's mouth with yours, and give two deep breaths; repeat the 30 + 2 cycles until the victim revives. When giving breaths, be sure that the victim's chest expands, and that the airway remains open. Continue giving CPR at the rate of 30 compressions followed by two breaths until medical assistance arrives. It is essential to keep oxygenated blood circulating as brain damage is only minutes away.

If the victim is unresponsive and an AED is available, unpack the unit, turn it on, and follow the automated voice instructions. This will require removing the victim's shirt, placing the pads on the chest as instructed, and letting the AED determine whether or when to give a shock. If you check for a pulse, do so at the wrist, not at the carotid artery. Blood pressure information for the rescue paramedics is valuable if the equipment is available in the first aid kit.

A shipboard first aid kit can be purchased at most marine supply stores, but can be easily made up in a waterproof tackle box. At a minimum it should contain the following:

Adhesive tape	Aluminum finger splint	Antibiotic ointment
Antiseptic solution	Antiseptic towelettes	Bandage strips
Breathing barrier	Cotton balls	Cotton-tipped swabs
Duct tape	Elastic wrap bandages	Examination gloves, nonlatex
Eye shield	Eyewash solution	Gauze rolls
Hand sanitizer	Instant cold packs	Nonstick sterile bandages
Petroleum jelly	Plastic bags	Safety pins
Scissors	Thermometer	Triangular bandage
Syringe	Turkey baster	Tweezers

In addition to the aforementioned equipment, certain medications should be in the kit, including aloe vera gel, antacids, antidiarrhea medication, antihistamine, calamine lotion, hydrocortisone cream, and pain relievers such as acetaminophen, aspirin, and ibuprofen. Always include a first aid manual.

Obtain as much information from a victim as possible and transmit that to the emergency medical technician. Use the acronym SAMPLE to populate your list: What are the Signs and symptoms, does the victim have Allergies, is the victim taking Medications, what is his/her Past medical history, what were the Last food and drink, and what were the Events leading to the crisis? As a "first responder" you are protected under Good Samaritan Laws, and are only expected to do your best to save a life. If you have not had a CPR course lately, put that on the agenda before going to sea!

Additional Reading

Bearden, B. 1990. *The Bluejacket's Manual*, 21st Edition. Annapolis: United States Naval Institute, 772 pp.

Exercises

1. You are the chief scientist aboard a UNOLS research vessel, and about to muster the scientific party for the first general meeting after leaving port. Make a list of the items to be covered in this conversation.
2. Become certified in CPR and first aid.
3. The following figure shows an instrument similar to one that may have been used on the *HMS Challenger*; it is about 30 inches long. What is it?

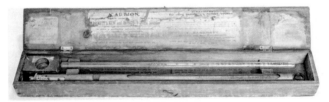

3

Research Vessel Construction—
Terminology, Equipment, and Machinery

Distinctive words identify parts of a ship, and they have an etymology dating back centuries and longer. The pointy-end of the ship is called the bow, not to be confused in pronunciation with "bow" as in bow and arrow. The other end of the ship is called the stern, and facing the bow, the right-hand side is called starboard, and the left-hand side is called port or port side. One might wonder, where did these terms originate? Language is a dynamic thing, and words change usage and meaning over time, but for the purpose of this chapter, a bit of wordcraft is in order.

Bow (pronounced "bough"), for example, seems to have been used in the fourteenth century and might be literally translated as the shoulders of a ship or perhaps from the bend in a bow (as in bow and arrow). Stern on the other hand, the hind part of a ship, might be related to steer, as in rudder, again probably Norse in origin from the thirteenth century. Starboard seems to have originated from the use of a steering oar near the stern on the right-hand side of a ship, the steer-board side; steering oars were known in Egypt five millennia ago. The opposite side, today's port side, probably is Middle English in origin derived from the term larboard, or loading side of a ship; it became the side facing the harbor (the port) in fifteenth century usage.

Figure 3.1 is a view from the starboard side of a proposed research vessel. The vessel particulars are molded length = 78 feet, waterline length = 73 feet, molded beam = 24 feet, and full load draft = 6 feet. The term "molded" refers to the maximum dimension, and molded beam (width) is usually measured amidships; a common term for molded length is length overall (LOA). Waterline length is horizontally measured from the intersection of the bow with the water to the stern, and draft is the vertical dimension from the waterline to the keel (the chief longitudinal structural member along the bottom from bow to stern). To scale, the person standing on the superstructure deck is 6 feet; he or she is standing on the fo'c'sle where the forecastle would be located on a fourteenth-century warship. The pilothouse, also called the bridge or sometimes the wheel house, is located just aft of the standing figure (see also Figure 3.4). On a larger ship, the bridge extends from port to starboard, houses all the navigation equipment, and may have bridge wings protruding out over the side for better visibility and navigation. The "roof" of the pilothouse is called the pilothouse deck, sometimes called the flying bridge.

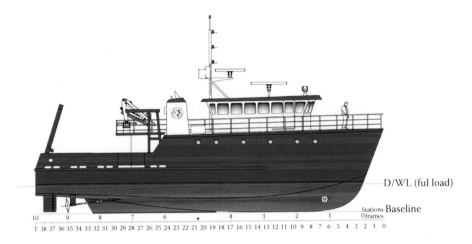

D/WL (ful load)

Stations Baseline
0 frames

10   9   8   7   6   •   4   3   2   1

T 38 37 36 35 34 33 32 31 30 29 28 27 26 25 24 23 22 21 20 19 18 17 16 15 14 13 12 11 10 9 8 7 6 5 4 3 2 1 0

FIGURE 3.1
Outboard profile of a 78-feet coastal research vessel for general nearshore oceanographic research (rendering by Boksa Marine Design for the Florida Institute of Oceanography). The round objects below the safety rail are portholes; the slot-like openings aft of the stack are freeing ports; the oval-shaped opening near the A-frame is a chock and is paired with the bits shown in Figure 3.4 as a fairlead for the mooring lines.

Terminology

Other features of the vessel in Figure 3.1 (from bow to stern): the guard rail is to protect one from falling overboard; the pilothouse has square-shaped windows, but the round structures below the superstructure deck are portholes. Above the pilothouse (fore to aft) are a search light, a radar antenna, and the mast with a second radar antenna, navigation lights, and a flag hoist (mast). Radar (an acronym for RAdio Detection And Ranging) is a very important navigational device, especially valuable in restricted visibility or fog. The navigational lights on the mast will be covered in Chapter 8, but briefly they tell other ships that the vessel is engaged in certain activities and cannot maneuver as required by the Rules of the Road. Similarly, the flags flown on the mast can inform other ships of the research vessel's activities, and can do so in any language using the International Code of Signals (Figure 3.2).

As your research vessel enters or leaves port, it may display the four-letter code that identifies its radio station. For example, if Whiskey-Foxtrot-India-Tango (WFIT) were flying in a vertical line, it would identify the ship as the *RV Miss Fit* (a mythical vessel no longer in service). The flags all have meanings, such as if Echo ("E" flag) were raised, it would mean that the ship is altering course to starboard. If on entering a port the "Q" (Quebec) flag were flying, it would mean that the ship is requesting pratique, that is, it is free of disease and wishes to clear quarantine and enter customs inspection. If the

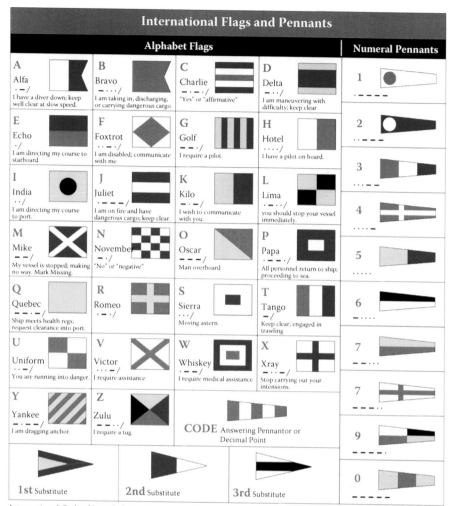

International Code of Signals flags and pennants from the National Geospatial and Intelligence Agency.

FIGURE 3.2

International Code of Signals flags and pennants from Publication 102, US National Imaging and Mapping Agency, 2003; now National Geospatial and Intelligence Agency (https://www. nga.mil). Numerous one-flag, two-flag, three-flag, and four-flag combinations are in Pub. 102, including place names, distress signals, and medical requests.

ship is visiting a foreign country, it may display as a courtesy the flag of the country being visited. Most often, the flag of the United States is flown from the mast shown in Figure 3.1 when the ship is underway.

Aft of the mast shown in Figure 3.1 are the stacks, that is, the engine exhaust structure, followed by a knuckle crane for lifting light items such as an outboard motorboat or cargo such as food, stores, and instruments. The crane is on a platform often called a catwalk. The superstructure deck is generally restricted to the ship's crew, but members of the scientific party can, with permission, visit. It is always courteous to ask permission to climb to this deck, and ALWAYS ask permission to enter the pilothouse.

The main deck as shown in Figure 3.1 (where the portholes are) has an open area toward the stern called the fantail. The waist-high structure enclosing the fantail is called the bulwark; it has freeing ports at ankle level to allow water to rapidly drain back into the sea. The large "A" frame for heavy lifting and instrument deployment is shown as the near-vertical structure above the transom (aftermost transverse frame). Most ships will require crew and passengers to wear hardhats when in this area, and may also require that all personnel working here wear a lifejacket too. This deck is called the main deck and has weather-tight doors that connect the fantail to the interior of the ship. These doors can be "dogged down" using handles bolted to both sides of the door. During good sea conditions, the weather-tight doors are usually kept open with a hook and eye, but during inclement weather they will be dogged down to prevent inadvertent flooding.

The interior and deck layout of the main deck are discussed in Chapter 5. Several other exterior features shown in Figure 3.1 include, near the bow and below the waterline, a "bow thruster" tunnel housing an impeller for maneuvering the ship, especially during docking and undocking. The aftermost structures are two rudders for directional control, which are just aft of the two propellers. This ship has twin propellers and is often spoken of as having "twin-screws" in nautical slang. Shafts connect the propellers to the engines, and struts stabilize the propeller–shaft system. Propellers are counter rotating, with the starboard one usually rotating clockwise when viewed from aft, and the port propeller counterclockwise when the engines are in the forward motion (ahead) positon. A stuffing box allows the shaft to rotate while preventing seawater from entering the engine room.

Structural Components

A ship of this size could be constructed of fiberglass, aluminum, steel, or some combination (e.g., steel hull and aluminum superstructure). In any case it would all start with the keel, and perhaps a keel-laying ceremony. The keel on such a vessel will be an interior "I beam" structure to reduce the draft and increase interior space. The bottom of the I is the keel plate, next the center vertical keel, capped by a rider plate at the top of the I; the rider plate often is at the height of the lowest deck, which is called the "A" deck, or inner bottom, or tank top if it covers spaces used for fuel or water. The fore-and-aft outer hull plate next to the keel plate is called the garboard strake, which is a holdover term from when ships were wooden plank construction. Figure 3.3 adds a few other structural member names such as the bilge strake, the floors (which are vertical transverse structures), the shear strake that defines the outer hull's shape, stringers running longitudinally, open pillars called stanchions to support the deck beams, and so forth. Ships have ceilings, but you walk on them; ships have girders, which run fore-and-aft, but not athwartships; ships have bulkheads, but no walls....

Seawater must be taken in to provide cooling water for engines, and this intake is through a structure known as a sea chest. In Figure 3.3, a cutaway of a sea chest is shown in the lower left-hand corner as a shaded compartment between the hull and the tank top. The intake hole through the hull is shown without a strainer for simplicity. A valve into the sea chest prevents flooding, and might be fitted with a thermistor for measuring intake temperature— which often is reported as sea surface temperature (SST), although it is not a traditional "bucket temperature" truly taken at the sea surface. The depth of intake temperature can vary widely depending on the ship's loading; this is not so problematic for a research vessel, but very much so for a tanker or a container ship.

Ships are compartmentalized into longitudinal blocks of space that can be isolated from each other in case of an accident or grounding. Watertight

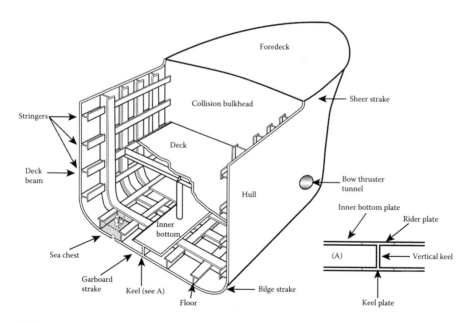

FIGURE 3.3

Hull construction of a welded steel ship in the size class of Figure 3.1. Many details are omitted for simplicity. The valve on the sea chest is called a sea cock, and is designed to prevent flooding.

bulkheads extend from the main deck to the tank tops. Doors through the watertight bulkheads, called watertight doors, can be closed remotely to prevent possible flooding from one compartment forward or aft into another. In case of emergency, these doors are closed and access to a compartment is vertical only. A small vessel is usually a one-compartment ship; that is, it will remain floating if one compartment is flooded. Larger ships may be two or more compartment ships, meaning two or more adjacent compartments may be flooded and the ship will remain afloat. It is important to know that during an emergency, escape from a compartment below the main deck is vertical only. Ships have ladders, but no stairs; fo'c'sle, but no forecastle; decks, but no roofs; overheads, but no ceilings....

The bulkhead shown in Figure 3.3 is watertight from the keel to the fo'c'sle. Such a watertight bulkhead would be called the collision bulkhead, and not be fitted with a watertight door, but rather with a bolted plate for access during maintenance. Forward of the collision bulkhead typically would be the (anchor) chain locker. On the fo'c'sle deck (Figure 3.4) is a piece of machinery called the anchor windlass, a specialized winch for hauling in the anchor chain. The anchor chain is fed from the windlass though the hawse pipe where it is bolted to the anchor. The windlass might also be fitted with one or two horizontally rotating drums called gypsy heads for hauling in mooring lines. If the ship had a bow observation chamber, it might be vertically accessed from this deck through a specially designed hatch set in a coaming (a raised frame to hold the hatch cover). The bow observation chamber might house a sea chest (see Figure 3.3) where water could flow through for sampling or routine monitoring with a thermistor for SST or salinity, and so on.

To complete the discussion on ship construction and general arrangement, Figure 3.5 shows the layout of spaces at the "A" deck level, that is, just above

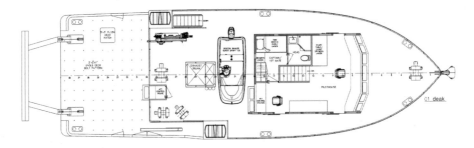

FIGURE 3.4
Deck arrangement for the superstructure deck (labeled 01 deck in drawing). The forward-most structure is the anchor chock. Machinery shown (fore to aft) are the anchor windlass, the knuckle crane on the port side amidships, and three hydrographic winches near the engine exhaust trunk, one on the centerline and two on the starboard side. The structures with two small circles are an overhead view of bits for securing mooring lines. The tubular objects, one on each side aft of the outboard motor boat, are self-inflating life rafts (rendering by Boksa Marine Design for the Florida Institute of Oceanography). LOA excludes the overhang of the A-frame.

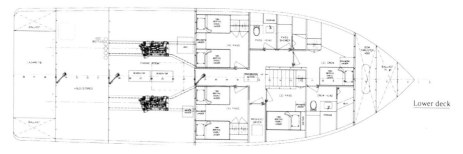

FIGURE 3.5
Machinery spaces and sleeping quarters for the ship shown in Figure 3.1 (rendering by Boksa Marine Design for the Florida Institute of Oceanography). Ship is driven by twin diesel engines port and starboard, and has two auxiliary generators shown amidships. Three watertight doors are shown connecting the engine room with the living quarters, the engine room with the hold, and the hold with the lazarette (the aftermost space).

the tank tops. Fore to aft, the chain locker and the bow thruster void are forward of the collision bulkhead, shown with a heavy line forward of the crew quarters. Other spaces forward of the engine room are the science party staterooms with over/under berths (ships do not have beds), lockers, and a chest of drawers, and crew quarters similarly outfitted. The head has a commode, sink, and a shower; there are no bathrooms on a ship. Passageways (there are no hallways) connect the staterooms to each other and to the ladder leading to the next deck above (discussed in Chapter 4).

Aft of the quarters area shown in Figure 3.5, through a watertight door, is the engine room or machinery spaces. This vessel is propelled by twin diesel engines, which are connected to the propeller shafts by a forward/reverse transmission. Also in the engine room there are two diesel generators for electric power. Proceeding aft through the next watertight door is a hold (storeroom) with CO_2 bottles that can be used to flood the engine room with gas in case of a fire. Farther aft is the last watertight door leading to the lazarette where the steering gear machinery would be located. The lazarette might also be used to store deck supplies such as mooring lines so as to keep them out of the weather when at sea. Machinery spaces are off limits to seagoing scientists and engineers as passengers, but may be visited with an escort from the crew under certain conditions. As with the pilothouse, ALWAYS ask permission to enter.

Additional Reading

Baker, E. 1953. *Introduction to Steel Shipbuilding*, 2nd Edition. New York, NY: McGraw-Hill, p. 398.

Exercises

1. Choose any University-National Oceanographic Laboratory System (UNOLS) Global Class Ship, download the vessel's deck plans (at least three levels), and identify the major structural components. Prepare a single-slide PowerPoint summary for presentation.

2. Choose any three words used in modern ship construction and research the etymology of them. Write three one-paragraph summaries of your research.

3. This 12″ device in brass and bronze is generically found on all ships. What is it?

4

Stability and Trim—Tonnage, Safety of Life at Sea, Maritime Organizations, and Sea Change

Research vessels essentially are special purpose self-propelled passenger ships (*RV Flip* is a clear exception). As such, loading of cargo, and the associated changes in draft and stability, are not major issues as they are with tankers, container ships, and freighters. Nevertheless, seagoing scientists and engineers should be aware of the considerations of the ship's officers regarding stability—the transverse ability of a ship to return to its original upright position after being inclined, and trim—the fore and aft inclination with respect to an even (level) keel.

As a ship moves in a seaway, it experiences six degrees of freedom, namely three rotational motions (roll, pitch, and yaw) and three linear motions (heave, sway, and surge). Pitch and roll are issues of longitudinal and transverse stability respectively, and yaw is rotation about a horizontal (x, y) plane usually in a following sea. Surge, sway, and heave are motions in the fore-and-aft, athwartships (side-to-side), and vertical directions, respectively. The motions are illustrated in Figure 4.1.

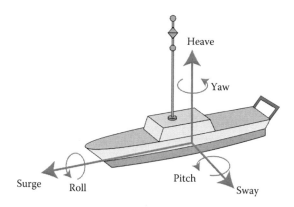

FIGURE 4.1

Ship motion in Cartesian coordinates, with x direction being fore-and-aft, y direction being from starboard-to-port, and the vertical (z) direction being opposite of gravity (positive upward). The ball-diamond-ball day shapes on the mast signify this is a research vessel that cannot maneuver due to the work in which it is engaged (see Chapter 8).

As a ship moves through sea and swell, it rolls with a period (T) dependent on the transverse stability. A tender ship is one that rolls gently, but with a long period of many seconds; a stiff ship rolls quickly and often in a jerking short-period motion. A stiff ship breaks all the dishes; a tender ship makes a sailor worry if she will capsize. The roll period is proportional to the square root of a transverse stability variable called the metacentric height, GM, where G is the center of gravity (mass) of the loaded ship, and M is the metacenter—a point to which G may rise and the ship will still possess positive transverse stability.

Transverse Stability

In Figure 4.2, the metacenter is seen as the intersection of two vertical lines one through B, the center of buoyancy when not inclined, and B', the buoyancy center at angle θ. M is determined by the naval architect from the ship's hydrostatic stability curves, and is provided to the ship's officers for loading and stability calculations. From the law of sines it is seen that $GZ = GM \sin θ$, and that for $θ \le 15°$ or so, $GM \approx KM - KG$. KB depends on the underwater shape of the immersed hull, and KG depends on the ship's loading and is cal-

culated using moments $KG = \sum (W_i \times VCG_i) / \sum W_i$, where W is the weight

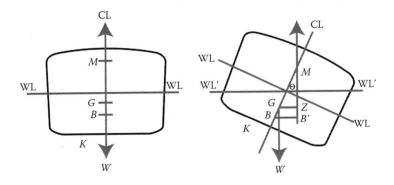

FIGURE 4.2

Transverse stability terms: M, metacenter; G, center of gravity; B, center of buoyancy when ship is upright; B', center of buoyancy when ship is inclined; K, keel; WL, waterline; CL, centerline; W, weight; and Z defines the righting arm GZ when the ship is inclined. The right-hand figure can be interpreted to show a ship listing to starboard, which could be a permanent condition and not necessarily one at risk of capsizing.

of an item (say a research submarine), and VCG is the vertical center of gravity of the item (say on the main deck aft). For many hull forms, $KB \cong 0.53 \times$ draft.

As an example of offloading a 10-ton research submarine and preparing it for deployment from a 500-ton research vessel, KG will change and so will GM, the stability parameter. The variables for the example problem are in the following table:

	Weights (tons)	VCG (feet)	Moments
Research vessel (Δ)	500	10	5000
Submarine	10	15	150
Difference (d)	490	5	4850

After offloading the submarine, $KG = \dfrac{4850}{490} = 9.9$ feet, and GM would be increased by 0.1 feet, that is, the ship is slightly more stable. If the interest is only in the change of G to say G' $\left(G \text{ to } G' = \dfrac{W \times d}{\Delta} = \dfrac{10 \times 5}{500} = 0.1 \text{ feet} \right)$.

The righting arm GZ is also now larger, and the ship is more able to return to an upright position. Figure 4.3 shows the operation in fine weather, but the seagoing scientist and engineer need to plan this operation carefully as weather changes quickly at sea.

FIGURE 4.3
Launch of the Johnson-Sea-Link I manned submersible (24 feet long; 11.5 tons) from the RV Edwin Link (168 feet long; 781 tons) of the Harbor Branch Oceanographic Institute. The ship's A-frame and tether management system were used to launch and recover the sub. Once in the water the sub was released from the cable and dived free to a maximum depth of 3000 feet. Photograph by John Reed, Harbor Branch Oceanographic Institute at Florida Atlantic University, used with permission.

The ship's roll period (*T*) is a function of the beam and the metacentric height, *GM*: $T = \dfrac{0.44 \times \text{beam}}{\sqrt{GM}}$, where 0.44 is an average roll-period parameter. For example, if a coastal research vessel has a beam of 30 feet and a metacentric height of 2.0 feet, the roll period would be about 9 seconds. For comparison, a typical 60-foot beam freighter with *GM* = 3.0 feet would have a roll period of 15 seconds; a tanker with a 70-foot beam and *GM* = 5.6 feet would have *T* = 13 seconds; and a passenger ship with an 80-foot beam and *GM* = 1.6 feet would have a very comfortable roll period of 28 seconds.

Tonnage

In the aforementioned example, the research vessel is listed as 500 tons. This is displacement tonnage given in long tons (2240 lbs per long ton), the mass of water displaced by the vessel. Oceanographers and ocean engineers usually think of mass in terms of density. Seawater density is typically cited as 1025 kg/m^3. To convert: 1025 kg/m^3 × 2.2046 lbs/kg × 1/(3.2808 ft/m^3) = 64 lbs/ft^3. Similarly, for fresh water the factor is 64 lbs/ft^3 × (1000/1025) = 62.43 lbs/ft^3, which is often rounded to 62.5 lbs/ft^3. The 500-ton vessel would displace: $500 \text{ tons} \times 2240\,(\text{lbs / ton}) \times \dfrac{1}{64 \text{ lbs / ft}^3} = 17{,}500 \text{ ft}^3$. For a typical research vessel, the volume displaced would be length × beam × draft × block coefficient, where the block coefficient for such a ship is about 0.58. A little arithmetic will show that this 500-ton ship could have dimensions of 100-foot long, 30-foot beam, and a 10-foot draft.

Two other tonnages are in common use: gross tonnage—space available to earn income, and net tonnage—gross tonnage minus spaces not available to carry cargo such as machinery spaces and living space. Gross tonnage is arbitrarily chosen to be the volume in cubic feet divided by 100. For boats, the following are the formulae where *L* = length, *b* = beam, and *D* = depth:

Term	Sailboats	Powerboats
Gross tonnage	$0.5\,L \times b \times D/100$	$0.67\,L \times b \times D/100$
Net tonnage	8/10 gross tonnage	9/10 gross tonnage

Both gross tonnage and net tonnage are space tonnage not displacement tonnage. Research vessels are not cargo ships, so although they may be given values for net and gross tonnage for taxation or canal transit cost calculations, these are not used in calculating stability.

Trim on a ship is a measure of longitudinal stability. At the forward perpendicular near the stem (the vertical line where the water line intersects the hull at the bow), and at the aft perpendicular near the stern (an imaginary vertical line through the rudder post), are permanent markings indicating how much draft there is. If the draft is in feet, the numerals are 6" high separated by a 6" space; if in metric units, the numerals are 1 decameter (10 cm) high separated by a 1 decameter space. If the readings fore and aft are the same, the ship is on an even keel. If not, then she is said to be "down by the head" or "down by the stern" as appropriate (sometimes referred to as trimming by the bow or by the stern). During a voyage, fuel and potable water are consumed, and the ship's buoyancy is altered. Keeping a ship in trim, that is, on an even keel, is an important responsibility of the chief mate as it maximizes longitudinal stability.

The ship's officers are trained in and tested on their required knowledge of stability and trim by US Coast Guard (USCG) inspectors. If your research cruise involves loading a heavy object on the ship, such as a research submarine or a portable laboratory van, the captain must be informed well in advance of the weight and desired location. Deck loads in general will reduce GM because the VCG of a van or submersible is typically on the main deck, well above G for the unloaded ship. The seagoing scientist or engineer can use the equations given earlier to make preliminary calculations and thus choose the vessel needed for the cruise.

Safety of Life at Sea

In April 1912, the Royal Mail Ship *Titanic* struck an iceberg southwest of Greenland, and sank quite rapidly. That tragic event led to the creation of two maritime organizations: the International Ice Patrol and the international convention that became known as SOLAS (safety of life at sea). The iceberg ripped open several adjacent compartments on *Titanic*, and she rapidly lost transverse stability due to flooding. Naval architects have offered numerous explanations for the materials failure leading to the sinking, but there is no explanation for the lack of sufficient lifeboats in the design (a *legal* capacity of 1178 in lifeboats for a maximum passengers and crew limit of 3327!).

Maritime history for the prevention of loss originally focused on the loss of cargo and income, not human life. Some of the earliest efforts date back to when Genoa and Venice were the dominant trading centers. There are scraps of evidence that Romans, Vikings, and the Hanseatic League paid some attention to the issue, primarily of cargo overloading, but the modern regulatory work is shared by France and the United Kingdom, particularly in the nineteenth century. The French and British efforts were independent, but not without communication. The rules these nations invented and promulgated primarily applied to their own ships, but it evolved into standards that ships of other nations visiting their ports would have to meet.

Then came April 14, 1912!

"Oh they built the ship Titanic, to sail the ocean blue. They thought they had a ship the water would never go through..." goes the song. The loss of life from the *Titanic* sinking was 1513 souls, precisely because the water did get through. Stunned owners and insurers, whose initial interest was financial, realized that such a horrific event should not have occurred, and that the cause might well be centered on human error: excessive speed in an iceberg field, technical arrogance as to the design, and lust for monetary gain. So it became the lot of the English, who lost most financially and in prestige, to call an international conference in London in January 1914, to address the prevention of such a catastrophe in the future.

The ship on which you will "sail" ("motor" actually; q.v. Figure 3.5) benefits in its design and construction and operation from the abysmal record of the loss of over a thousand UK ships a year—yes over 1,000 per year—in the late nineteenth century. More importantly though is the move toward international standards for ships and the safety of their crew, regardless of the flag of the country under which they operate. "Flags of Convenience" were the response by many corporations, that is, registering and licensing in countries that were not signatories to such onerous rules. Eventually, the international standards from London in 1914 and three subsequent conferences led to the creation of the International Maritime Organization (IMO), an agency of the United Nations, and SOLAS. Most maritime nations are IMO members.

SOLAS has embraced satellite technology for search and rescue (SAR). Modern research vessels are equipped with emergency position indicating radio beacons (EPIRBs), which are about the size of a shoebox. All modern EPIRBs operate on 406 MHz and send signals that can be received by geostationary operational environmental satellites (GOESs) as well as polar-orbiting satellites. EPIRBs with Global Positioning System (GPS) receivers, sometimes called GPIRBs, are much preferred as GOES can identify not only the vessel but also its latitude and longitude (polar orbiting satellites depend on Doppler tracking and can locate in the 2–5 km accuracy range whereas GPS is in the 2–5 m range). EPIRBs should be either manually activated or automatic; automatic activation occurs when the water depth is 4 m. In the United States, Coast Guard SAR teams are trained to receive EPIRB reports via GOES and commence operations usually within 2 hours or less.

Personal locator beacons (PLB) also operate on the 406 MHz international emergency frequency, but are smaller (about the size of a cellphone) with less powerful transmitters. A PLB is easily integrated into a life vest, and of course can be used on land as well as at sea. Considering that PLBs cost less than a few hundred US$, and EPIRBs or better yet GPIRBs less than a few hundred US$ more, no credible research organization should allow students and staff to leave port without them, nor without training in how to use them properly.

The (ship) automatic identification system (AIS) is a vessel-to-vessel or vessel-to-shore SOLAS radio beacon transceiver method of identifying a ship, and broadcasting its position, course, and speed to others. It operates at 161.975 and 162.025 MHz, so its is range limited to 10 or 20 nautical miles.

AIS-equipped ships, and all of 300 gross tons or more are so required, transmit every few seconds data essential to prevent collisions. These data are usually displayed on Electronic Chart Display and Information System (ECDIS) computers, which are approved by the IMO as a replacement for traditional paper nautical charts. Smaller vessels may not be AIS equipped, so the prudent mariner still needs to be cognizant of the traditional rules (Chapter 8) for the prevention of collisions (COLREGS).

Your research vessel, especially in the United States when operated by the University-National Oceanographic Laboratory System (UNOLS), National Oceanic and Atmospheric Administration, or US Coast Guard, has over 100% capacity in the life rafts and at least one EPIRB. Each person has a life preserver, a bunk card, undergoes weekly safety drills, and has a USCG licensed or commissioned crew on a ship inspected regularly. As a seagoing scientist or engineer, your serious attention to instructions and training is a life-or-death exercise. IMO rules are primarily designed for commercial vessels, but the legacy of Samuel Plimsoll, a Member of Parliament in 1876, with great political cost for his courage, legislated regulations that ships could not overload and thus reduce the freeboard (vertical distance) necessary to provide sufficient buoyancy for expected weather conditions. Plimsoll marks (Figure 4.4) are the legacy of his willingness to politically sacrifice for the welfare of others.

Other US Maritime Organizations

The American Bureau of Shipping (ABS) (AB in Figure 4.4) is a classification society; they are a nonprofit organization of engineers and naval architects

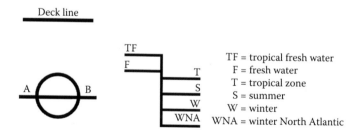

FIGURE 4.4
Plimsoll marks from the American Bureau of Shipping. The horizontal line through the circle is the maximum draft to which a ship may be loaded for summer temperate conditions. When more freeboard is required, the ship may be loaded only to the designated line such as winter (W) or winter North Atlantic (WNA). Since a vessel will float higher in salt water than freshwater, the freeboard may be reduced to tropical fresh water (TF) or fresh water (F) as appropriate. Freeboard is the vertical distance between the deck line and the load line.

that create and enforce regulations as to the safe design and construction of merchant vessels. The US Maritime Administration (MARAD) is a US government agency in the Department of Transportation; they work with the ABS and ship operators among others to regulate and grow the industry. The USCG is headquartered in the US Department of Homeland Security; although the Coast Guard has many activities such as SAR and law enforcement, they play a vital role in marine transportation by licensing deck and engine officers, issuing qualification certificates for seamen, and inspecting vessels to insure that they meet standards of safety and stability. The Society of Naval Architects and Marine Engineers (SNAME) is a professional society dedicated to education and career development; national SNAME meetings bring together members of the ABS, MARAD, USCG, and others to share progress in engineering and design of all classes of vessels.

The International Research Ship Operators (IRSO.org) is a voluntary group of organizations with the common interest of promoting safe, efficient, and environmentally responsible research vessel operation, and coordinating international ocean science and engineering activities. They meet annually, have no budget (on purpose), and exchange information and schedules so as to better collaborate projects and cruises. The IRSO works with the European Research Vessels Operators, the US Research Vessel Operators Committee, and the Ocean Facilities Exchange Group. The IRSO has five themes: (1) research vessel builds, modifications, and performance; (2) manning, safety, and training; (3) scientific technology; (4) legal and insurance; and (5) cooperation and outreach. Membership is open to research vessel operator institutions and commercial interests.

Sea Change

Commander Locke Cranford was the chief marine engineer (CME) on the US Coast and Geodetic Survey (USC&GS) Ship *Explorer*. He and I (Figure 4.5) ate many meals together in the officer's wardroom (commissioned officers' mess), where after dinner the chief yeoman would show a movie (yes there really were 16-mm films once!). All members of the crew were invited into the wardroom to enjoy the movie, but Commander Cranford was always half paying attention to the film. He was listening to the engines and the sound of the waves slapping against the hull, and timing the roll period, the creaking and groaning of the superstructure, and the flickering of the lights, as the *USC&GSS Explorer* moved with the sea. He was listening for a sea change.

Fast forward two decades, and as a chief scientist the lessons of CME Cranford are well appreciated. Listen to your ship! Does the sound of the oceanographic winch change? Has the ship changed course on time? Is the next oceanographic station at 1600 or 1630? What does 1630 mean to the scientific party? Do they understand the 24-hour method of time keeping? Is (e.g.,) the CTD ready for the upcoming station? Is the wind/sea a safe operating environment for them? Does the captain, mate, or officer on duty know what the next oceanographic station is meant to measure? Is it safe? Has there been a sea change?

The chief scientist is a listener. Listening to the sounds of the sea against the hull; listening to the scientific party; listening to the crew, who after all, are most familiar with the way this particular ship behaves in a seaway. There is no separation between the crew and the scientific party. Respect the skills of all members of the ship's party: licensed or commissioned officers, enlisted or noncommissioned crew, and newbies making their first voyage as an oceanographer or ocean engineer. All are to be respected for their knowledge, experience, and passion. Harken back to CME Cranford: "feel your ship."

To close the loop, recall First Lieutenant George Riser (Figure 2.1), not to forget CME Cranford, Chief Yeoman Ralph Fortna, Chief Survey Technician J.D. Lewis … heroes all. Names forgotten to most, but to a very inexperienced USC&GS Ensign (Figure 4.5), taught to sense a "sea change," extraordinary teachers all. Going to sea is a very personal experience. As members of the scientific party and especially as chief scientist (if you are privileged to be so assigned), respect and listen to the sea: your ship, your crew, and your colleagues. A "sea change" can be more than winds, and waves, and currents.

FIGURE 4.5

Ensign George A. Maul, US Coast and Geodetic Survey (USC&GS), inspecting a Roberts Radio Current Meter aboard the USC&GS Ship *Explorer*. (Photo cgs01581, National Oceanic and Atmospheric Administration's Historic Coast & Geodetic Survey Collection.)

Additional Reading

George, W.E. 2005. *Stability and Trim for the Ship's Officer*, 4th Edition. Baltimore, MD: Cornell Maritime Press, p. 528.

Exercises

1. Your 3-ton portable laboratory van will fit on the fantail of a 100-ton research vessel that you'd like to charter. What will be the change in GM if the metacentric height originally is 4.0 feet, the ship's VCG = 8 feet, and the van will be 13 feet above the keel? Is this a good ship to charter?

2. Dr. Bill Richardson (1924–1975) and Mr. Bob Charnell (1937–1978), both fine seagoing scientists, both friends, both lost when their research vessels sank: Bill in the Gulf of Maine and Bob in the Molokai Channel. Search for a serious accident to a research vessel, write a one-page summary of your findings, and prepare a one-slide PowerPoint of the incident. What is the role of maritime agencies in such a tragedy?

3. Naval architects at one time actually hand drew plan and profile views of their creations, in part using "what is it?"

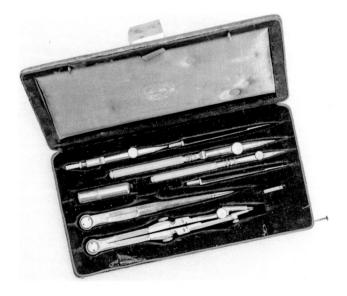

5

Science Spaces—Open Deck, Wet Lab, Dry Lab, Computer Lab, Stateroom, Lockers

Science spaces on ships before the *RV Albatross* (Figure 1.1) were converted from the original design, usually military, to new usage. Clearly the "Great Cabin," if it was so named on the *HM Bark Endeavor* of Captain James Cook, was a "laboratory." His many original surveys added mightily to Great Britain's cartographic knowledge of the sea, and using the newly invented chronometer by John Harrison, added significant improvement to the longitude of the many headlands and bays he charted. To put this into perspective, Cook's first voyage was contemporary with the stirrings of independence of the original 13 American colonies in North America *ca.* 1769–1771.

Science spaces on a modern research vessel fall into several categories: open deck, semienclosed, enclosed, and hull penetrating. This is not meant to exclude the navigation spaces (q.v. Figure 3.4), because knowing where and when a sample is taken is as important as the data in the sample itself. The point of "where/when" cannot be overemphasized (and will be developed further in Chapter 6); if the seagoing scientist or engineer does not know the geographic position (latitude [ϕ] or longitude [λ]), depth (z), and time (t) of a sample, there is little value in the sample!

Deck Layout

In Figure 5.1, the layout of the main deck of the 78-foot length overall (LOA) coastal research vessel shown in Figure 3.1 is rendered. From fore to aft: chain locker, store room, mess (workboats do not have dining halls), galley to starboard (ships do not have kitchens), ship's office to port where the chief steward (Figure 2.3) keeps accounts and so on, and into the combination wet/dry lab.

The "dry lab" is an air conditioned and heated space where sensitive chemical instruments such as an autoanalyzer or salinometer are operated. As shown in Figure 5.1, the "wet lab" space is the starboard side of the wet/dry lab. Aft on the open deck is where water samples, geologic grabs, or biological critters are transferred from the underwater instruments into sealed containers for analysis. On larger research vessels, the wet lab would be a separate semienclosed space; Figure 5.1 shows that the sample collection area is on the afterdeck, somewhat protected from the weather.

FIGURE 5.1
Layout of main deck on same 78-foot length overall vessel discussed in Chapter 3. (Courtesy of Boksa Marine Design for the Florida Institute of Oceanography.)

The "dots" drawn on the afterdeck, also spoken of as the fantail, are flush bolts in the University-National Oceanographic Laboratory System pattern (2 feet by 2-feet centers and 1-in.-diameter bolts) that allow standard interchange of portable instruments or machinery such as a specialized winch (see Figure 11.1). The aftermost structure is the "A-frame" from which the hydrographic wire would be suspended; the A-frame can be hydraulically rotated ±30° or so from the vertical to facilitate handling instruments and lifting heavy objects onto the fantail. A snatch block and/or a meter wheel (Figure 5.2) would be suspended from the cross bar of the A-frame with a shackle.

Parts of a Meter Wheel

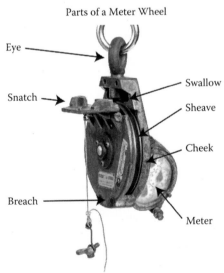

FIGURE 5.2
A snatch block allows wire or line to be inserted through the cheek (side) instead of feeding through the swallow (upper opening above the sheave). Most meter wheels are of the snatch block design, where the sheave must be matched to the wire diameter for accurate measurement. The meter wheel pictured is about 12 in. high; the sheave is designed for 1/8 in. (3.75 mm) hydrographic cable.

Computers are in common use on modern research vessels, and would be placed in the dry lab area shown in Figure 5.1. On a large research vessel, the computers may be in a separate room to prevent any splashing that may occur when a sample is analyzed using electrochemistry. It is critical that any electronics be isolated from water, and that they be securely fastened to a bench to prevent damage in rough seas. Modern wet/dry labs will have frequency-controlled 60 Hz 110 V alternating current (AC) power (US standard), and tie-down points. Savvy seagoing electronics technicians are absolutely essential members of the scientific party.

Additional specialized science spaces could include a bow observation chamber, a moon pool, and instrumented sea chests. A bow observation chamber would be located below the waterline in a bulbous bow structure and have several viewing ports. A moon pool is a vertical shaft, like an elevator shaft, from the main deck through the hull; it is used for lowering instruments and for divers to enter and exit the ship without having to go over the side. Ships have sea chests in many locations for the purpose of having a controlled connection through the hull; if the sea chest is instrumented, it can provide a continuous supply of near-surface data such as temperature and salinity. A sea chest is usually built between floors, from the tank top to the hull (see Figure 3.3).

The scientific party will require storage space for samples taken, and for supplies needed during the voyage. Usually, there is assigned or designated space for this purpose, but if not, an estimate of that needed must be made. An inventory of supplies should be included in the project instructions (see Chapter 13). It must be ascertained that sufficient space of proper conditions are available. The scientific party is responsible to properly secure their stores; a bit of marlinspike seamanship skill (see Chapter 9) will be essential.

Other Spaces for the Scientific Party

On the open deck, the fantail shown in Figure 5.1, or the foredeck shown in Figure 3.1, the seagoing scientist or engineer is a "seaman." It may well be that you will need to wear a hard hat, steel-toed shoes, a life vest, and other such protective gear. Do so! While you may be handling an instrument or remotely operated vehicle (ROV) worth hundreds of thousands of US$, here you are a seaman. You may need to very quickly grab a line and "make it fast." You may be the reason a fellow seaman did not get washed overboard. The deck is slippery; the ship is rolling, surging, pitching; something is "fouled" (a line jammed in a block). Your environment is exciting and dangerous. Remember the five P's: proper preparation prevents poor performance....

The scientific party will have staterooms. In Figure 3.5, a stateroom is shown as being shared by two people each with their own bunk. Some staterooms have four bunks; some more. Usually, each member of the scientific party will have their own bunk as "hot bunking" (sharing a bunk) is not particularly pleasant. Lockers (ships do not have closets) for hanging clothing may also be shared as are drawers in a chest; space is at a premium. Most likely there will be a communal head with or without a shower. A head is drawn as shown in Figure 5.1 on the port side just forward of the wet/dry lab; keep it clean and sanitary—this is shared space.

On a small research vessel such as shown in Figures 3.1 and 5.1, the mess is both a place to eat and a place to "veg out" between watches or stations. On larger ships (q.v. Figures 1.1 and 5.3) there may be several eating facilities: an officer's mess—called the wardroom on a ship with commissioned officers (q.v. Figure 2.2), a captain's mess, a chief's mess for chief petty officers (CPOs), and a crew's mess for other noncommissioned officers. A UNOLS ship (q.v. Figure 2.3) may have a common mess for all personnel, but there may be a table set aside for the captain, the chief engineer, and the chief scientist. Do not presume that you can sit down anywhere; ask if it is permissible before doing so. Common seagoing courtesy is to recognize the line-of-command, and to respect it. The chief scientist will assign you to a specific mess for the duration of your voyage. ALWAYS remove your cap when entering the cabin or mess.

On a large National Oceanic and Atmospheric Administration ship or a US Coast Guard cutter, the commanding officer will have a cabin, and often a separate mess. Usually, the term "cabin" is reserved for the private space of the CO of such a vessel, and may contain his/her stateroom, dayroom, and sometimes a captain's galley with his/her own steward. The CO's cabin is always in close proximity to the bridge and may have a voice tube for communication. This is a very formal setting, and members of the crew and scientific party do not have access unless specifically invited. As with the bridge, engine room, galley, chief's (CPO's) mess, crew's mess, and CO's cabin, or any stateroom not your own, always ask permission to enter.

Some spaces in the ship will be restricted from the scientific party. This is for your safety and for smooth operations. Perhaps the best mindset is for members of the scientific party to behave as if they were guests. You will make new friends, learn new skills, and mature in respect for the ways of the sea. At least six millennia of experience goes before you. Respect that as hard earned experience for survival on Earth's most perilous surface.

Large Research Vessel Science Spaces

On a modern large research vessel, by which is meant 238-feet LOA or so as pictured in Figure 5.3, many of the shared spaces seen in Figure 5.1 are separated. For example, there will be a wet lab and a dry lab. The basic structure discussed in Chapter 4 will remain: bulkheads, watertight doors, sea chests, heads, hull, weather deck, pilothouse, overhead, and so forth. A ship of this size (Figure 5.3) may have a bow observation chamber, a weather-balloon shack, a gravity lab, or an acoustic instruments lab, all of which will be under the supervision of survey technicians and the chief scientist. A few examples follow.

Aft of the bridge on a large ship, especially a hydrographic survey ship, will be the plotting room. This space is the center for gathering bathymetric data. If the ship is conducting swath bathymetry, then control of the courses steered will be commanded from this room. The bridge officer will still be the authority for safety, and may countermand a proposed course or speed based on the immediate situation, say a ship crossing the bow from starboard to port (see Chapter 8). The plotting room will have fathometers, a Global Positioning System (GPS), gyrocompass repeaters, computers, and most likely a plotter so that the progress can be visualized.

On the same deck there can be a radio room and a chart room. A separate chart room is necessary so that at night the officer on watch does not lose "night vision," and the lighting is only from red light bulbs. The radio room will be the center for communications back and forth to the home office, and where a licensed radio operator maintains watch. A small head, using again only red light, will be in this complex on the bridge deck. Navigation for ordinary operations will be on the bridge itself, and most likely there will be a captain's chair; everybody else "stands" their watch to stay alert.

FIGURE 5.3
Ocean class research vessel Sally Ride (AGOR-28), built by Dakota Creek Industries, Inc. of Anacortes, WA. RV Sally Ride was delivered to the US Navy in 2016. (Photograph used with permission of Dakota Creek Industries.)

A bow observation chamber, if in the design, is forward of the collision bulkhead, and is accessible by vertical ladders through several decks. Each deck will have a watertight hatch to isolate the chamber, deck by deck. The chamber itself will have several viewing ports, electrical power, communications equipment, and may be the location of a sea chest for scientific instruments to measure sea surface temperature, sea surface salinity, chlorophyll-a, and so on. This is not a science space for a claustrophobe, but viewing reefs, fish, and porpoises swimming by is an exciting and memorable experience.

If a ship has a gravity meter, which is also known as gravimeter, the "gravity lab" will be just above the tank tops amidships. This is the part of the ship where vibrations and accelerations due to heave and roll and surge (see Figure 4.1) are least, and where the gravimeter's accelerometers are better able to measure Earth's gravity field. Seaborne gravimetry is best conducted on submarines, but some useful data have been acquired by surface ships with a specially designed science space.

Larger research ships will have a separate computer lab to isolate the electronics from airborne sea salts. Often too, the lab will have additional cooling as well as controlled frequency AC power. Shipboard computer systems gather data from a variety of data-intensive instruments such as acoustic Doppler current profilers (ADCPs), sound navigation and ranging (SONAR) systems of various purposes and designs, and possibly weather RADAR, to name a few. In addition, the computers are logging navigation information, time, voluntary observing ship (VOS) surface weather observations (see Chapter 12), and of course the chief scientist's notes.

Research vessels such as shown in Figure 5.3 (Armstrong-class ships), are built by the US Navy. They are assigned to major US oceanographic institutions engaged in naval research. The *RV Sally Ride* (AGOR-28) is at the Scripps Institution of Oceanography, and the sister (brother?) ship, the *RV Neil Armstrong* (AGOR-27), is at the Woods Hole Oceanographic Institution. While there are larger ships, these vessels are extremely capable of more than a month at sea, and a cruising range in excess of 10,000 nm, while accommodating 24 scientists and 20 crew. They are the vanguard of the next generation of research vessels owned by the US Navy and operated by institutions.

Additional Reading

Hayler W.B. 1989. *Merchant Marine Officer's Handbook*, 5th Edition. Atglen, PA: Cornell Maritime Press, Schiffer Publishing, p. 589.

Exercises

1. GPS is the modern standard of electronic navigation. Research this method of trilateration (the term "triangulation" is often misused as the methodology), and write a one-page summary of this means of satellite technology. Why are a minimum of four satellites required for three-dimensional (3D) positioning (show the equations)?

2. You are standing on the flying bridge as your ship is preparing to get underway. Your ship is lying portside to. Sketch the pier, the hull, and name the following: bow line, stern line, breast line, forward quarter spring line, after bow spring line, bollards, bits, chocks, and cleats. With the wind blowing from starboard to port, what was the order of letting go the lines? Why?

3. Dr. Ross Austin was an inventive Scripps Institution of Oceanography oceanographer, and developed this 18-in.-long device. What is it?

6

Mariner's Compass—True and Magnetic, Variation, Deviation, Correcting, and Uncorrecting

The origin of the magnetic compass is not definitive, but probably came to Europe from China where it was used for maritime direction finding *ca.* 1117. In 1190, Alexander Neckam wrote of the use of the magnetic compass for traversing the English Channel. One hundred years or so later, the dry magnetic compass is described, an event contemporary with the development of Portolan Charts in *ca.* 1290. Early scholars credited Flavio Gioja (*ca.* 1302) with the mariner's compass as is known today—combining a compass needle and the compass card, but his existence is in doubt. There is a statue of Flavio in Amalfi, Italy—so there is a monument to someone who did not exist who did not invent the compass?

The science behind the compass, geomagnetism, was beginning to be appreciated with *De Magnete*, the work of William Gilbert in 1600. Gilbert thought that Earth's magnetic field was fixed, but Henry Gellibrand in 1635 published a treatise showing that from actual measurements the geomagnetic field is indeed changing. In 1701, Sir Edmond Halley (for whom Halley's Comet is named), published a map of Earth's magnetic field *variation*, that is the difference between true north and magnetic north (Figure 6.1). Halley is also credited with inventing isogonic lines, which are lines of equal magnetic variation (geophysicists call variation "magnetic declination," but herein the term "declination" is reserved for navigational astronomy—the angle north or south of the celestial equator of a star or the sun or moon or planets; Chapter 15). Halley was a contemporary of Sir Isaac Newton, published papers in astronomy and in meteorology, and added to the development of the diving bell.

Halley's General Chart of the Variation of the Compass, published *ca.* 1701 (Figure 6.1), laid the foundation for magnetic compass information on charts. Modern nautical charts show a compass rose with the difference between true north and magnetic north, the rate of magnetic variation change, and the date upon which the information is based. A stylized example of the compass rose on a nautical chart is shown in Figure 6.2. The compass rose is likened to a wind rose, and was probably developed in Italy during the 1300s; it is contemporary with Portolan Charts (Appendix A).

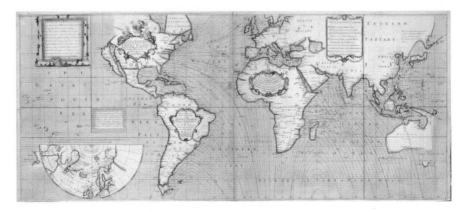

FIGURE 6.1
Reproduction of Halley's geomagnetic map, first presented to the Royal Society in London in 1700, but printed *ca.* 1701. The map is based on several years of voyaging by Halley throughout the Atlantic Ocean and other sources. The Lionel Pincus and Princess Firyal Map Division, the New York Public Library.

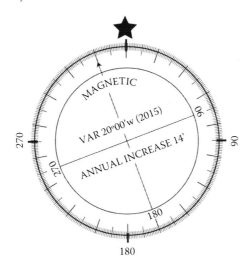

FIGURE 6.2
Compass rose as seen on a nautical chart. The magnetic variation shown is 20°W in the year 2015, and the variation is increasing 14 arc minutes per year. The cardinal points of a compass are north, east, south, and west; the intercardinal points are northeast, southeast, southwest, and northwest. "Boxing the Compass" involves learning all 32 points (11 1/4° per point).

Marine Compasses

Modern research vessels will have magnetic compasses as well as a gyroscopic compass, often called the "gyro." Electromechanical gyrocompasses have a spinning disk and require electricity to operate. They are designed to point toward true north and maintain that direction through gyroscopic precession. Gyros are not affected by the ship's ferrous materials, but are subject to errors due to the ship's motion and friction. These too must be checked by the navigator, who often will compare the gyro with the magnetic compass to assure that the gyro is functioning properly. Many traditional navigators will observe the sun's amplitude at sunrise and sunset (see Chapter 15) to check the gyro twice daily. Optical and laser gyros are now developed. Most ship's officers will compare compasses to the Global Positioning System (GPS) computed direction as the ship moves (requires forward motion), yet if there is a power outage, the magnetic compass is still the working standard.

Magnetic compasses aboard ship are influenced by two major factors: Earth's magnetism and the effect of the ship's structure on the compass. As noted, the first effect is called magnetic variation by mariners, which is defined as the horizontal angle that a perfect magnetic compass differs from true direction to the north geographic pole. The second effect, which is of the vessel itself, is referred to as the magnetic deviation. Thus in using the magnetic compass, the terms are true (T), the direction toward the north pole; variation (V), the effect of the geomagnetic field; magnetic (M), the direction toward the magnetic pole of a perfect compass; and deviation (D), the difference between a perfect magnetic compass and the actual compass (C). Mariners learn the mnemonic can dead men vote twice (CDMVT) to apply compass corrections.

Using Figure 6.2, if a mariner wished to steer true north (000°), he or she would need to steer 020° magnetic (M) plus any other correction needed for deviation (D). Converting from T to M is known as *uncorrecting*, and it should be obvious that adding a westerly variation is necessary to convert from T to M. Similarly, when uncorrecting an easterly variation, the mariner would subtract the variation. The same arithmetic applies for uncorrecting from M to C: uncorrecting add west (W) or subtract east (E). If in the case of Figure 6.2 the deviation when the ship is heading true north is (say) 10°E, then T→V→M→D→C would be 000°T + 20°W = 020°M–10°E = 010°C—that is, on this particular ship and heading, the mariner would steer 010° by the ship's magnetic compass (C) in the wheelhouse.

The Napier Diagram

Deviation of the magnetic compass on a research vessel depends on the course upon which the vessel is headed and the cargo she may be carrying. To determine the magnetic deviation, a process known as *swinging ship* is employed. With today's modern ship equipped with GPS and a gyroscopic compass, determining the deviation table is much simplified. The process is to steer the ship at normal cruising speed in a series of true directions, say 000°, 015°, 030°, and so forth and read the ship's magnetic compass. The readings are recorded on a form and/or plotted on a graph known as Napier's Diagram. For each true heading, the magnetic variation is added or subtracted to obtain the magnetic heading, and the deviation is calculated from the compass reading at that true direction. An example of such a form is given in Figure 6.3.

In the T→V→M→D→C example given earlier, the entries would be in the first data-entry row, and read left to right: $000° + 20° = 020°-10° = 010°$. That is, for this particular vessel for this particular loading, in this particular geographic region, the helmsman would have to steer $010°$ on the pilothouse magnetic compass to make $000°$ true. Note that the deviation depends on the ship's heading, so if the magnetic compass is used to take a bearing (direction from the ship toward a shore object), the deviation correction depends on the ship's heading, not the bearing. For example, if the bearing to a lighthouse is (say) $270°$ by the ship's compass, the true bearing to the object C→D→M→V→T would be $270° + 10°E = 280°M-20°W = 260°T$. This procedure is called *correcting* (going from compass to true) and the signs of deviation and variation inform the navigator whether to add or subtract. Correcting Add East (CAE) is often memorized to assist in this process. So for this example, the navigator would plot the bearing as a line of position to the lighthouse of $260°$ on the nautical chart; that is, the ship is somewhere along a line drawn at the true angle to a meridian of $260°$ through the lighthouse. Position fixing by visual bearings is discussed in detail in Chapter 7.

		Deviation table			West deviation			East deviation		
TRUE	Variation	Magnetic	Deviation	Compass	$-20°$	$-10°$	$-0°$	$+10°$	$+20°$	
000										
015										
030										
045										
060										
075										
090										
105										
120										
135										
150										
165										
180										
195										
210										
225										
240										
255										
270										
285										
300										
315										
330										
345										
360										

FIGURE 6.3
Combined deviation table (left-hand side) and Napier's Diagram (right-hand side). An alternate form would have the courses $000°$, $015°$, $030°$, and so forth in the column labeled magnetic, thus giving a Napier's Diagram centered on magnetic north.

The Binnacle

The magnetic compass on a ship is housed in a structure called the *binnacle* (Figure 6.4). Often a ship will have two such binnacles, one in the pilothouse, and one directly above the pilothouse on the deck above (called the flying bridge). Binnacles are complicated instruments that house internal magnets and two large iron spheres athwartships of the compass called quadrantal spheres. The compass is suspended on *gimbals* to allow it to remain level while the vessel moves with the sea. A vertical line called the *lubber's line* is set fore-and-aft, parallel to the keel. A person steering the ship, called the *helmsman*, reads the direction that the ship is heading by visually aligning the lubber's line with the compass card.

FIGURE 6.4

A binnacle from a Liberty Ship (World War II transport). The unit stands about 5 feet high, weighs about 200 pounds, and is on a teak base. The compass is housed in the brass hood, and is lighted by a kerosene lamp (upper right). The quadrantal spheres are painted red and green, port and starboard, respectively. A sighting vane is the topmost round brass fixture on the hood, used for taking a bearing to an object. Compensating magnets are in the stand, and the Flinders Bar, which houses soft iron as needed, is behind the stand (not shown).

A marine magnetic compass is housed in a compass bowl that is filled with oil to dampen the motion of the ship. The compass bowl is gimballed to keep it level, and can be directly fitted with an azimuth circle (sighting vane). If an azimuth circle is used, the hood shown in Figure 6.4 is removed and the instrument is set directly on the compass bowl's outer ring. The compass shown in Figure 6.5 is about 10.5 inches in diameter, and is a precision navigation instrument, perhaps the single most important instrument aboard.

Steering a ship, whether by magnetic compass or gyrocompass, is an art best learned from practice. The central tenat is that the ship moves about the compass, not the compass rotating as the ship moves in a seaway. In the aforementioned example, that is steering 010° by the ship's magnetic compass (C), if the helmsman sees the ship is heading (say) 005°, then a small amount of right rudder is necessary to return the ship to 010°. Conversely, if the helmsman notices that the ship is heading (say) 020°, then a small amount of left rudder is needed. If the helmsman applies the wrong rudder angle, the ship will veer farther off course, a situation called *chasing the compass*. Once on course, it is always easier to look ahead and steer toward an object—say a point of land or a cloud (for a short time of course). Modern ships have self-steering mechanisms, but they are not used in *heavy weather* or when entering or leaving port as there are many instances where vessel traffic requires the helmsman to steer clear of another ship or obstacle. The savvy watch officer always checks to see that the *autopilot* and the compass agree, as mechanical failure is possible.

Adjusting the compass is a complicated process, and few mariners today have the necessary skills. Thus, this short introduction to the ship's magnetic compass is greatly simplified. Many magnetic fields develop on a ship, not only from the mass of the vessel itself but also from other electrical and

FIGURE 6.5
Mariner's compass housed in the binnacle shown in Figure 6.4. The azimuth circle would fit on the outer brass ring, which is 10.5 inches in diameter. Note the compass is mounted in hinged gimbal rings (both visible) to keep the instrument level in a seaway. The top of the Flinders Bar tube is seen at the center top of the photograph. The compass rose shows all 32 points.

electronic devices such as motors. Even a pocketknife can disturb the magnetic compass. On a long voyage, the magnetic variation (V) is changing (Figure 6.1), and the compass course steered must be regularly adjusted or the ship will not stay on the true (T) course desired. The seagoing scientist or engineer is encouraged to visit the pilothouse to learn more about the history and practice of navigating using the magnetic compass, but be sure to ask permission to do so from the officer on watch so as not to interfere with his or her other work.

Additional Reading

National Geospatial-Intelligence Agency. 2004. *Handbook of Magnetic Compass Adjustment..*, Formally Publication No. 226. Bethesda, MD: National Geospatial-Intelligence Agency, 45 pp.

Exercises

1. The *RV Delphinus* has used GPS to "swing ship" and determine the compass deviation in the offing of Cape Canaveral. Earth's magnetic variation here is 3°45'W (1988—annual increase 9'), and there is no set and drift of the boat due to wind or currents (i.e., the GPS course made good is the true heading with respect to the geographic north pole). In the following table, determine the true course; determine compass deviation and plot it out on a Napier Diagram.

GPS Course	Variation	Magnetic Course	Deviation	Compass Course
–	–	000	–	010
–	–	030	–	045
–	–	060	–	080
–	–	090	–	115
–	–	120	–	130
–	–	150	–	150
–	–	180	–	175
–	–	210	–	200
–	–	240	–	225
–	–	270	–	265
–	–	300	–	305
–	–	330	–	337
–	–	360	–	010

2. You plot the following true courses and distances on a chart made with a Mercator projection. Calculate the three compass courses that you need to steer on the *RV Delphinus* in order to reach the points in your trip from the sea buoy off Fort Pierce Harbor to the Gulf Stream and return: (1) 090° T for 30 nm, (2) 180°T for 40 nm, (3) Calculate return true course and distance; calculate compass course to return (you will need to interpolate deviations from the earlier table).

3. Earth's magnetic field is central to the design of this Ekman-era instrument. What is it and how is geomagnetism used?

7

Coastal Navigation—Nautical Charts,
Geographic Positioning, Marine
Electronics, and Instruments

Ships have been plying the ocean for millennia, and they always need a means of determining their position (geographic location). Most scholars believe that early sailors travelled along the coasts, keeping familiar landmarks in sight. Arguably the ablest ancient mariners were the Phoenicians (*ca.* 1200 BC), who spread their culture throughout the Mediterranean Sea and beyond. Thales of Miletus (*ca.* 585 BC) invented the gnomonic chart projection, and brought navigational astronomy to the Greeks from the Phoenicians; he thought that Earth is a sphere. Sea charts as they are commonly known seem to be the invention of the Genoese (see Appendix A). The oldest extant sea chart is the *Pisan Chart*, dating only to *ca.* 1290 AD, and perhaps this is because maritime knowledge was considered a "trade secret."

The modern nautical chart uses the Mercator projection, invented by Gerardus Mercator in 1569. Mercator's projection has the advantage that meridians of longitude are orthogonal to parallels of latitude, and that sightings (visual bearings) and compass courses plot as straight lines. A Mercator chart is a cylindrical projection with the cylinder tangent at the equator, and with the origin of the projection at Earth's center (Figure 7.1). A unit of longitude, say $\Delta\lambda = 1°$, on a Mercator projection is equal to one degree of latitude $\Delta\phi = 1°$ only at the equator. For example, if on a Mercator chart, 1 inch $= \Delta\lambda = 1°$ of longitude at the equator, then 1 inch $= \Delta\phi = 1° \times \sec\phi$ at latitude ϕ. That is, convergence of the meridians on a sphere is a function of the cosine $\left(\dfrac{1}{\sec}\right)$ of latitude.

The distance in nautical miles on a Mercator projection is given by the difference in latitude. By definition, the nautical mile is 1 arc-minute of latitude, and thus there are $90° \times 60\dfrac{\text{nm}}{°} = 5400$ nm from the equator to the pole. Also by definition, the great circle distance from the equator to the pole is 10,000 km, and thus there are $\dfrac{10,000\text{ km}}{5,400\text{ nm}} = 1.8519\dfrac{\text{km}}{\text{nm}}$; a great circle is the

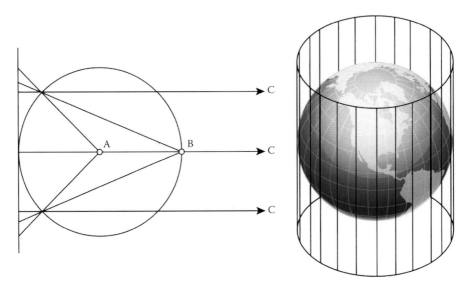

FIGURE 7.1
Left panel: Azimuthal chart projections—A, gnomonic; B, stereographic; and C, orthographic.
Right panel: Mercator's cylindrical projection. In Mercator's projection for nautical use, the
cylinder is tangent at the equator; oblique and transverse Mercator projections are also in use
but rarely for maritime applications. (Redrawn from Publication No. 9, National Imagery and
Mapping Agency, Bethesda, MD; originally written by Nathaniel Bowditch as the *American
Practical Navigator*, 1802).

line on a sphere formed by the intersection of a sphere and a plane through
the sphere's center. Thus, one degree of latitude is $60 \text{ nm} \times 1.8519 \frac{\text{km}}{\text{nm}}$ or
111.1 km. In 1929, an international convention established the nautical mile
as 1852 m exactly or 6076 feet.

   Since a chart is a flat representation of a sphere and meridians converge
toward the poles, in Cartesian coördinates a north–south distance
$\Delta y = 111.1 \frac{\text{km}}{\circ} \times \Delta \phi(\circ)$, and an east–west distance $x = 111.1 \frac{\text{km}}{\circ} \times \Delta \lambda(\circ) \cos\phi$. For
example, on a perfect sphere at $\phi = 30°$ the actual north–south distance of 1 arc-
minute of latitude $\Delta y = 111.1 \frac{\text{km}}{\circ} \times \frac{1}{60}° = 1.852 \text{ km}$, and the actual east–west
distance of 1 arc-minute of longitude $\Delta x = 111.1 \frac{\text{km}}{\circ} \times \frac{1}{60}° \times \cos 30° = 1.604 \text{ km}$;
on Earth, which is an oblate spheroid, these relationships vary slightly and
can be found in Table 7 (length of a degree of latitude and longitude) in *The
American Practical Navigator* (Pub. No. 9, US Defense Mapping Agency).

The Nautical Chart

The scale of a nautical chart (Figure 7.2) is the ratio of the distance on a chart divided by the distance on Earth. Thus, a small-scale chart (say 1:1,000,000) covers a large area, and a large-scale chart (say 1:5,000) covers a small area, such as a harbor. For example, if the scale is 1:5000, then 1 in. on the chart is 5000 in.

$$\left(5000 \text{ in.} \times \frac{1}{12} \frac{\text{feet}}{\text{in.}} \times \frac{1}{6076} \frac{\text{nm}}{\text{feet}} = 0.0686 \text{ nm} = 0.0686 \text{ arc-minute of latitude} \right)$$

on Earth. The scale is chosen at the mid-latitude of the chart on a Mercator projection, because the scale varies with latitude. The scale and other pertinent information is printed in the chart's legend, usually placed on a land area not critical for marine navigation information. All details of US nautical charts are in *U.S. Chart No. 1* (http://www.nauticalcharts.noaa.gov/mcd/chartno1.htm), available free online, and published by the Office of Coast Survey, National Ocean Service (NOS) of the National Oceanic and Atmospheric Administration (NOAA).

FIGURE 7.2

National Oceanic and Atmospheric Administration Nautical Chart No. 11481 showing the approaches to Port Canaveral, Florida. The scale of this chart is 1:25,000 at $\phi = 28°25'$N latitude in the print edition. As printed in this book the scale is changed because the size of the paper on which it is printed has changed. Soundings are in feet.

In marine navigation, speed is given in nautical miles per hour or knots (kn). The term "knot" probably came from Bartolomeu Crescêncio of Portugal, at the beginning of the sixteenth century. A line with a chip of wood attached was tossed overboard and the sailor would count the number of knots (spaced every 47.25 feet) passing through his hands in the 28 sec a sandglass would

$$empty \left(\frac{47.25 \text{ feet}}{28 \text{ sec}} \times 3600 \frac{\text{sec}}{\text{hr}} \cong 6076 \frac{\text{feet}}{\text{hr}} = 1 \text{ kn} \right). \text{ The chip-log method gives}$$

an estimate of the ship's speed through the water, but not speed over the ground. Oceanographers often work in meters per second ($m \cdot s^{-1}$ or m/s) when measuring current speeds, but wind speeds are still mostly stated in knots. For handy reference $1 \, m \cdot s^{-1} = 1.94 \text{ kn} = 2.24 \text{ mph} = 3.6 \text{ kph} = 3.28 \text{ fps}$, where mph is statute miles (5280 feet) per hour, kph is kilometers per hour, and fps is feet per second.

All charts and maps have datums (in tidal usage the plural of datum is not data but datums). The datum information is printed in the chart legend (Chart 11481, Figure 7.2, uses the North American Datum of 1983 for the projection). Water depths on a chart are called soundings and for the United States are either in feet or fathoms (6 feet = 1 ftm). Obviously it is important to know which units are used, but the fathom has been used for millennia, and was mentioned by Posidonius in 150 BC who reported a sounding of over 1,000 ftm. The vertical distance for soundings was originally made with a leadline—a "rope" with a lead weight at the end. The soundings datum on modern NOAA charts is mean lower low water (MLLW). Earlier NOAA charts of the east coast United States used mean low water (MLW) as the datum for soundings, so again the chart legend needs to be consulted. MLLW and MLW refer to the value of the low tides every 24.84 hours averaged over an epoch of 19 years (because the lunar nodal tide's period is 18.61 years).

Elevations on a nautical chart (Figure 7.3) are given in feet above mean high water (MHW). The intersection of MHW with the juxtaposed land is the *shoreline*, a legal boundary that often demarks private property from public property (the intersection of MLW with the coast is the *baseline*, which is used to define territorial seas such as the 12-mile limit). While soundings use MLLW as their datum as a safety measure, heights such as bridge clearances use MHW also for safe-passage reasons. Lighthouse elevations too are reported in feet, which makes them handy for determining a position. For example (Figure 7.2), the height (H) of Florida's Cape Canaveral Lighthouse is $H = 137$ feet, and if a navigator using a sextant measures the elevation angle to be (say) $a = 1°$ then, in the absence of refraction and with the observer being at sea level

$h = 0$ feet, since $\tan 1° = \dfrac{\text{opposite}}{\text{adjacent}} = \dfrac{137}{\text{distance}}$, distance = 1.3 nm (the actual

equation is $d = \sqrt{(\tan^2 a / 0.0002419^2) + ((H - h)/0.7349)} - (\tan a / 0.0002419)$).

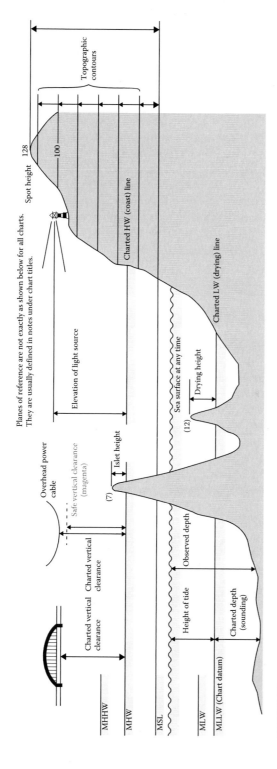

FIGURE 7.3

Elevations and depths on a nautical chart are related to mean tidal heights averaged over a 19-year epoch. A tide gauge is placed in the vicinity of a chart, and measurements are analyzed for the mean lower low water, mean low water, mean sea level, mean high water, and mean higher high water. Nineteen years is chosen as a convenience because the lunar nodal cycle is 18.61 years.

(From Nautical Chart No. 1.)

Now if a compass bearing is also taken from the ship to the lighthouse, both the direction and distance to Cape Canaveral Lighthouse are known. The navigator can now plot his/her position on the chart. The process is known as taking a visual fix, and is one of several methods used in practice.

Position determination by visual means, such as the aforementioned Cape Canaveral Lighthouse example, is an essential skill in the absence of electronic navigation availability. The required equipment is a chart and chart table, a compass with which to take bearings (direction from the vessel to the object on land), plotting equipment (parallel ruler, straight edge, dividers, and pencil), and if horizontal sextant angles are to be taken, a sextant and a three-arm protractor. Three examples are given in Figure 7.4: position A from simultaneous bearings of two shore-side objects; position B from two simultaneous horizontal sextant angles of three shore-side objects; and position C from a bearing and distance to a single shore-side object (the aforementioned Cape Canaveral Lighthouse example).

Visual Navigation

The process of visual position fixing and navigation is known as piloting, and stems from the word "pilot", which means an especially skilled mariner working in a local area. Pilots were known to the Ancient Greeks and the Romans, and the word is used in very early manuscripts such as Homer's *The Iliad, ca.* eighth century BC. Piloting is more than position fixing; it includes estimating a vessel's future position, by projecting the course and speed on a chart, a process called dead reckoning (DR). The navigator always needs to know where the ship will be so as to avoid danger, where to make turns, or where to stop to pick up a harbor pilot.

The vessel shown in Figure 7.4 is on course 145° true, or southeast. It passes numerous navigational aids on its passage. The northernmost navigational aid is labeled "Fl 10 sec 159 feet 16M" which identifies a lighthouse that is 159 feet tall, can be seen 16 nm at sea (assuming the observer's height of eye is $h = 15$ feet), and has a white light that flashes every 10 sec. To the south-southeast along the coast the next navigational aid is a microwave tower, followed by a tank, and lastly "Gp FL(3) 60 sec 137 feet 26M RBn 313 – – · ·" which is a group of three closely spaced white flashes that repeats itself every 60 sec, is 137 feet tall with a light that can be seen 26 nm at sea, and has a radio beacon transmitting on 313 kHz, sending the Morse code letter "Z" (– – · ·). The point on the ship's course marked "DR 0700" is the 7:00 AM dead reckoning position estimated from a prior fix, by time and speed along the course line.

At position A in Figure 7.4, the navigator obtains a fix by two simultaneous bearings (240° and 287°, respectively) with respect to true north on the microwave tower and the 159-foot lighthouse. At position B, the navigator takes

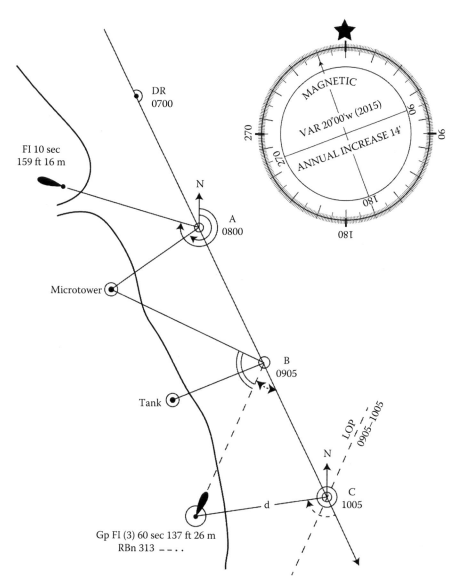

FIGURE 7.4

Visual position fixing or piloting along an imaginary coast where the magnetic variation is 20°W and increasing 14 arc-minutes per year. Position A is from two visual bearings, position B is from two simultaneous horizontal sextant angles, and position C is from range and bearing to the lighthouse. Ship's course is 145° T at 12 kn.

two simultaneous horizontal sextant angles by using the same center object (the tank) for the left angle (group flashing light to the tank) and the right angle (tank to microwave tower); using a three-arm protractor, the angles are set and when all three arms of the protractor intersect the three objects, the position is fixed. Finally at C, the fix is, as discussed earlier, knowing the lighthouse is 137 feet tall and the vertical sextant angle is 1°, coupled with a bearing of 260° T on the lighthouse.

The number to the lower right of each fix is the time of the fix: 0700 for the DR position, 0800 for A, 0905 for B, and 1005 for position C. Recall that ships use a 24-hour time system with four numerals to designate hour and minute (Chapter 2), thus 10:05 AM is written as 1005 (and say 1:45 PM is written 1345); the term "o'clock" is never used at sea. Navigating along a coast, or *coasting*, requires regular fixes to avoid hazards such as shoals, rocks, and wrecks. Keeping track of time and position is an essential habit for safety of life and property. The coast in Figure 7.4 is imaginary, but gives examples of aids to navigation, and the practice of visual sightings to determine position.

If the navigator cannot obtain a definitive fix, it may be necessary to obtain a running fix. At position B in Figure 7.4, if the direction of the dashed line were also observed, it forms a line of position (LOP) that can be advanced along the ship's track according to the ship's course and speed. Then if at a later time (positon C in Figure 7.4), another bearing is taken to the same lighthouse, the intersection of the 1005 lighthouse bearing with the advanced 0905–1005 LOP is a *running fix*. Alternately, notice that the relative bearing to the lighthouse from B and then to the lighthouse from C has doubled; then from plane trigonometry, the distance run from B to C is equal to the distance off at C, a process called *doubling the angle on the bow*. The doubled angle in Figure 7.4 is 45°–90°, but other combinations such as 30°–60° or 22 1/2°–45° give the same result: distance run between bearings is equal to distance off at the second bearing [in a plane triangle doubling the angle on the bow results in the angle at B and the angle at the lighthouse (LH) being equal, and from $\frac{\sin B}{d} = \frac{\sin LH}{BC}$ the distance off (d) is calculable]. In any case, a running fix is never to be as trusted as a definitive fix.

Electronic Navigation

Position "C" could also have been obtained by RADAR; alternatively, the direction only could have been obtained with a radio direction finder (RDF). RADAR was developed during World War II, and today is standard equipment on most research vessels. RADAR sends and receives radio waves and creates an image of the surrounding land, waters, buoys, and other ships. It is quantitative and provides both direction and distance to targets, and if a lighthouse is in the RADAR image, the vessel's position can be determined. RDF on the other hand will provide a bearing to any radio source, but not distance. In the case of position C in Figure 7.4, the receiver on the ship is tuned to

313 kHz, the Morse code letter Z is identified (see Figure 3.2), and a bearing is obtained. Modern RDF units may have automatic bearing circuits; older units will require more navigator skill and time.

Bearings may be true, relative, or magnetic. True bearings are illustrated in Figure 7.4, and in practice could have been taken from the gyroscopic compass repeater on the wing of the bridge using an azimuth circle (see Chapter 6). If the ship is not equipped with a gyrocompass, the bearing could be magnetic, again using an azimuth circle fitted over the magnetic compass or a handheld device. Magnetic bearings would have to be corrected for the local magnetic variation and ship's deviation as written in Chapter 6. Once converted to true bearings, the magnetic bearings can be plotted as shown in Figure 7.4. Relative bearings are those taken with respect to the ship's bow, the bow always being 000°, abeam to starboard being 090°, dead astern being 180°, abeam to port being 270°, and so forth (see Figure 8.1). The ship's heading, taken at the instant of observing a relative bearing, must be added together to obtain a true or magnetic bearing to the object (true° = relative° + heading°).

Modern large ships will carry Electronic Chart Display and Information System (ECDIS) computers. ECDIS uses digital charts, either Electronic Navigation Charts (ENC) or Digital Nautical Charts (DNC) in the program; ENCs are vector representations of paper chart data, and DNCs are raster (scanned) representations. If either method of chart presentation meets International Maritime Organization (IMO) specifications, it satisfies the safety of life at sea (SOLAS) (see Chapter 4) agreements to which IMO member nations ascribe. Positioning in ECDIS is most often from Global Positioning System (GPS), and in most circumstances is ± several meters to several tens of meters in accuracy. Along the coasts of the United States, the US Coast Guard operates differential GPS (dGPS) technology, and dGPS positioning is ± 1–2 m, good enough for nautical chart surveying.

The curious seagoing scientist and engineer should visit the navigation deck on his or her voyage. Ask permission from the officer on watch to come into the bridge or pilothouse, and enter into a discussion about the navigation system in use on that particular vessel. Most ship's officers are delighted to share their experience and love of the sea with visitors when navigation duties are not too hectic, such as when they are entering or leaving port or in fog. The same is true of the ship's engineering spaces; visit with the engineering officer on watch, again asking permission to enter the control room beforehand.

Additional Reading

Bowditch, N. 2003. The American Practical Navigator, 2002 Edition. Pub. No. 9, U.S. Government Printing Office, 873 pp. Available online at: http://trove.nla.gov. au/version/42085784.

Exercises

1. You will be planning an oceanographic voyage from Palm Beach to Miami to Bimini to Great Isaac Light to Freeport to West End and back to Palm Beach. Obtain the NOAA nautical chart of the area and lay out the cruise. Your ship is the same *RV Delphinus* from Chapter 6; she has a cruising speed of 7 kn and draws 6 feet of water. Pencil each leg of the voyage, determine the true courses, calculate the compass courses, and calculate the travel times of each leg in hours. What instruments will you need to determine direction and distance?

2. The pressure gradient force per unit mass (pgf) at the sea surface is given by $g\dfrac{\partial h}{\partial x}$ and $g\dfrac{\partial h}{\partial y}$ in the east–west (x) direction and the north–south (y) direction, respectively, where g is the acceleration of gravity and h is the sea surface height. Your captain reports that h has changed by +1 m as the ship sailed from $\phi = 36°N$, $\lambda = 75°W$ to $\phi = 35°N$, $\lambda = 74°W$. Calculate the pgf between 36/75 and 35/74.

3. This device is about 9 in. in overall length. What is it?

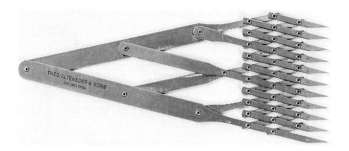

THEO. ALTENEDER & SONS

8

Rules of the Road—Classes of Vessels, Lights, Day Shapes, Maneuvering, and Collision Avoidance

Shipmasters, owners, passengers, and crew have all been interested in avoiding collisions at sea, yet it was not until 1889 in Washington DC that an international conference was convened to codify the steering rules and lights and day shapes and signals that ships need to employ. Of course there is a long history of setting such rules in place: England deserves much of the credit for the thought process that eventually led to 1889 and the creation of COLREGS, the *International Regulations for Preventing Collisions at Sea*. Currently 72 COLREGS, a result of the 1972 International Maritime Organization led convention, is the definitive source of the rules.

Relative direction from one vessel to another is the foundation of collision avoidance. No matter what geographic course a ship is steering, the relative directions are the same. Dead ahead is 000° relative, not necessarily true north. Dead astern is 180° relative, 090° relative is broad on the starboard beam, and 270° relative is broad on the port beam. Intercardinal points are 045° relative is broad on the starboard bow, 135° relative is broad on the starboard quarter, 225° relative is broad on the port quarter, and 315° is broad on the port bow. Details of the relative directions are shown in Figure 8.1. Knowing relative directions is an important communication tool; for example, if someone falls overboard, the direction to that person will be clearly understood using these terms (see Chapters 2 and 7).

There are 32 relative directions and a bit of arithmetic shows that each point is $(360°/32 \text{ points}) = 11\frac{1}{4}°$. Running lights on ships are specified in the 72 COLREGS in points. For example, the starboard sidelight is green and shines from dead ahead (000°) to two points abaft the starboard beam $22\frac{1}{2}°$ past 090° or an arc of $90° + 22\frac{1}{2}° = 112\frac{1}{2}°$ from dead ahead. Similarly, the red port sidelight shines from dead ahead to two points abaft the port beam; same 112.5° arc, but in the opposite direction. The stern light, which is white, shines from dead astern forward to two points abaft the starboard beam and two points abaft the port beam, an arc of 12 points or 135°. Thus, a ship's sidelights and stern light cover an arc of 112.5° + 135° + 112.5° or 360°, fully around the horizon.

For the United States, there are actually two sets of navigation rules: those for international waters and those for inland waters. The 72 COLREGS

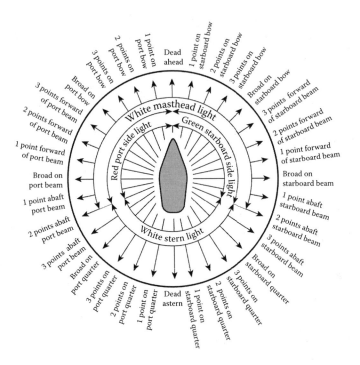

FIGURE 8.1

Relative bearings using classical 32-point nautical terminology, and showing the arc of visibility of the masthead light (20 points), the stern light (12 points), and the sidelights (10 points each). Range of visibility in nautical miles varies and is dependent on the size of the ship.

Demarcation Line is shown on nautical charts (q.v. Figure 7.2), and is usually across channels at or near the shoreline. A vessel properly lighted for international waters (q.v. Figure 8.1) can legally operate in US inland waters, but a vessel lighted strictly for inland waters may not operate in international waters using those lights. The US Inland Navigation Rules Act of 1980 did much to homogenize the two sets of rules (international and inland). The current US Coast Guard publication on the rules is printed so that the international rule is on the left-hand page and the corresponding inland rule on the right. This layout is very effective for learning this important aspect of Admiralty Law.

Seagoing scientists and engineers voyaging on a professionally staffed ship will have little need, other than curiosity, to know the Rules of the Road. However, if you operate a small craft, then knowledge of definitions, and lights, shapes, signals, steering, and so on are mandatory—this is not an option! In this chapter, the presentation will be an overview that hopefully leads to thorough reading and learning extramurally.

The international rules apply to all vessels on the high seas and on all waters connected thereto; inland rules apply to all US harbors, rivers, and the Great Lakes. Nothing in the rules exonerates vessels from the consequences of neglect of any precaution required by the ordinary practice of seamen, and due regard must be had to all dangers, which may make a departure from the rules necessary to avoid collision. Admiralty Courts have held owners, crew, and the master liable for this central responsibility.

A few of the general definitions are listed as follows:

- A "vessel" includes nondisplacement boats, seaplanes, wing-in-ground (WIG) craft, and anything else capable of being a means of transportation on water.
- A "power-driven" vessel is one being propelled by machinery, and a "sailing vessel" is one being propelled by sail alone even if it has machinery not being used.
- A "vessel engaged in fishing" excludes fishing by trolling (using lines only); a research vessel towing large sampling nets is fishing.
- A "vessel not under command" is a vessel unable to maneuver as required by the rules, and is unable to keep out of the way of others.
- A "vessel restricted in her ability to maneuver" means that because of the nature of her work she cannot keep out of the way as required; a research vessel with hundreds of meters of hydro wire over the A-frame falls into this category.
- A "vessel constrained by her draft" is a power driven vessel unable to maneuver outside of restricted waters such as in a narrow channel.
- The word "underway" means a vessel is not at anchor, or aground, or made fast to the shore; underway does not imply "making way" through the water.
- Vessels are in sight of one another when one can be **visually** observed by the other; note the RADAR limitation.
- "Restricted visibility" includes fog, mist, snow, heavy rain, sandstorms, and so on.
- Lights are required from sunset to sunrise and in restricted visibility; day shapes are required from sunrise to sunset.

Collision Avoidance

The requirements for maneuvering to avoid collision are called the "steering and sailing rules." There are three sections: (1) Conduct of vessels in any condition of visibility. (2) Conduct of vessels in sight of one another. (3) Conduct of vessels in restricted visibility. Vessels, when in sight of one another (2), even if it is foggy or misty, have rules different than in (3) restricted visibility, that is, when another ship might be heard, but not sighted visually. A "give-way vessel" must take action to avoid collision; a "stand-on vessel" must hold course and speed if safely practicable.

Regarding (1): All vessels must have at all times a proper "lookout," a qualified person who by sight and hearing can make a full appraisal of the situation. A vessel must at all times maintain a "safe speed," that is, so she can take effective action to avoid collision in the operating conditions. "Risk of collision" exists if the relative bearing of another vessel does not change. For example in Figure 8.1, if the relative bearing of another ship is "broad on the starboard bow" and remains constant, there is risk of collision. Action to avoid collision shall be positive, made in ample time, and with due regard to good seamanship. In narrow channels, anchoring and fishing are not permitted, and vessels in the channel should stay to the right-hand side; traffic separation schemes exist with rules similar to channels.

Concerning item (2): Sailing vessels when in sight of one another on the port tack keep out of the way of a sailing vessel on the starboard tack (Figure 8.2a), that is, imagine which sidelight would be seen—if red keep out of the way (q.v. Figure 8.1). When on the same tack, the sailing vessel to windward keeps out of the way of the leeward sailboat. A sailing vessel running free (i.e., wind abaft the beam) keeps out of the way of a sailing vessel close hauled (i.e., wind forward of the beam).

Any vessel overtaking another (Figure 8.2b) keeps out of the way; that is, imagine if seeing the white stern light and not sidelights, keep out of the way; generally alter course to port when overtaking. Power-driven vessels meeting head to head (Figure 8.2c), that is, each seeing both sidelights, must alter course to starboard to avoid collision. When power-driven vessels' courses cross, the vessel seeing another on her own starboard side must give way (again imagine the sidelight color—if red, give way, Figure 8.2d).

In general, never alter course to port to avoid collision except when overtaking, and make it clear that you as the give-way vessel are making a substantial change—either altering course to starboard or slowing or backing down. All vessels must give way to a vessel not under command, a fishing vessel, or a vessel restricted in her ability to maneuver (such as a research vessel engaged in work where, for example, she cannot engage her engines and start making way).

Regarding (3): "Restricted visibility" means that vessels are not in sight of one another, and that they must proceed at a safe speed. "Safe speed" has been held by Admiralty Courts to mean that your ship can stop in half the distance of visibility. If a close-quarters situation is detected by RADAR, action to avoid collision must be taken early, never altering course to port. If the fog signal of another vessel is heard forward of the beam, speed must be reduced to the minimum needed for steering until danger of collision is over. Fog signals of vessels engaged in various occupations will be covered in the section on sound signals.

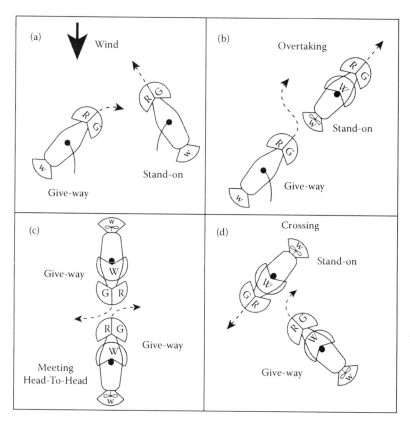

FIGURE 8.2

Steering and sailing rules: (a) Two sailboats crossing. (b) Any vessel overtaking another (sailboat illustrated) is a give-way vessel. (c) Two power-driven vessels meeting head-on or nearly head-on. (d) Two power-driven vessels crossing. Power-driven vessels shown with 225° white (W) masthead light and a propeller; sailboats do not show a masthead light unless propelled by both sail and machinery. Note that lights are shown from sunset to sunrise and in reduced visibility.

Lights and Shapes

More detail on lights and shapes are shown in Figure 8.3. Note that this is a very truncated version of all the combinations of lights and shapes, but for a research vessel under 50 m in length (defined in the rules as length overall [LOA]; q.v. Figure 3.4), it captures the most common ones. Lights and shapes for international waters and for inland waters are mostly the same, but not exactly. The 72 COLREGS define lights in degrees instead of relative bearings as shown in Figure 8.1, but they are all based on the traditional wording. Masthead lights, the stern light, and special occupation lights as shown in Figure 8.3 must be placed over the ship's centerline; sidelights are placed so as best to mark the molded beam, green to starboard and red to port.

"All-round light" means a light showing an unbroken light over an arc of the horizon of 360°. If a vessel such as an air-cushion vessel operating in nondisplacement mode shows an amber "flashing light," it requires a frequency of 120 flashes or more per minute.

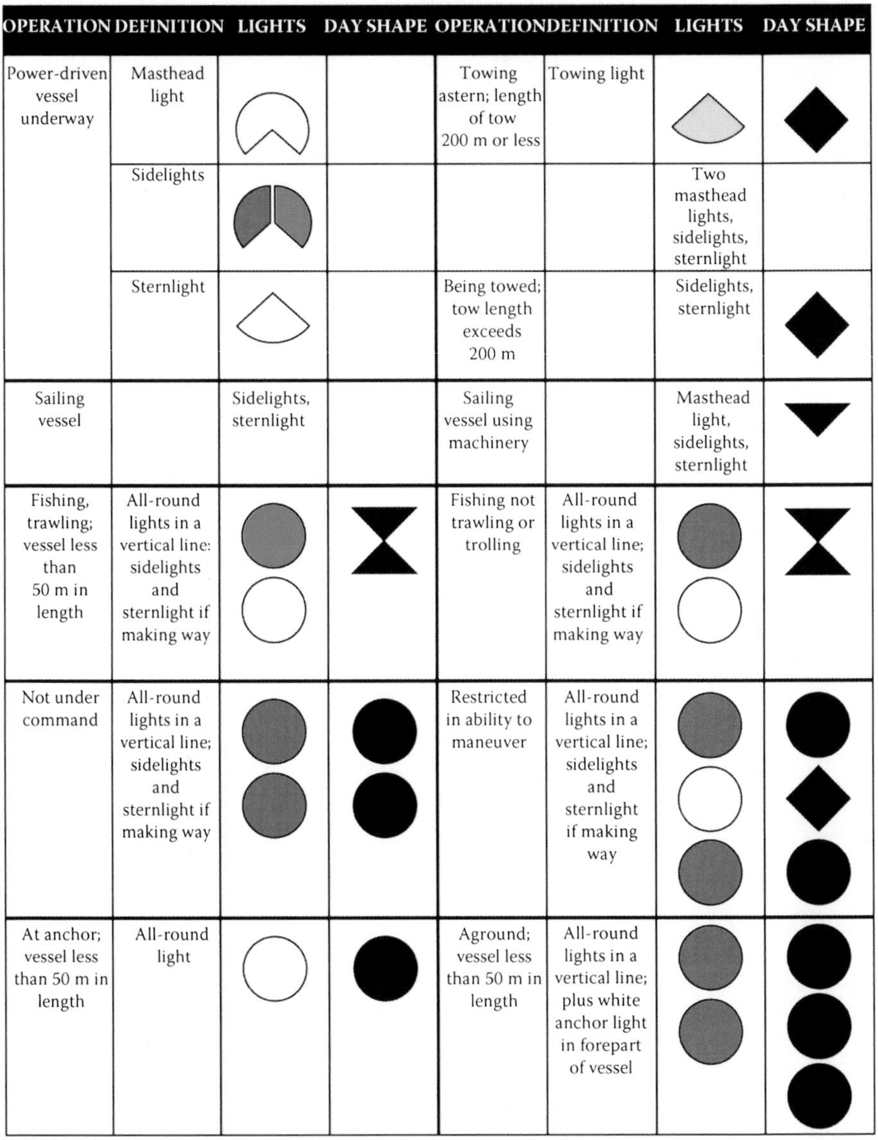

OPERATION	DEFINITION	LIGHTS	DAY SHAPE	OPERATION	DEFINITION	LIGHTS	DAY SHAPE
Power-driven vessel underway	Masthead light			Towing astern; length of tow 200 m or less	Towing light		
	Sidelights				Two masthead lights, sidelights, sternlight		
	Sternlight			Being towed; tow length exceeds 200 m		Sidelights, sternlight	
Sailing vessel		Sidelights, sternlight		Sailing vessel using machinery		Masthead light, sidelights, sternlight	
Fishing, trawling; vessel less than 50 m in length	All-round lights in a vertical line: sidelights and sternlight if making way			Fishing not trawling or trolling	All-round lights in a vertical line; sidelights and sternlight if making way		
Not under command	All-round lights in a vertical line; sidelights and sternlight if making way			Restricted in ability to maneuver	All-round lights in a vertical line; sidelights and sternlight if making way		
At anchor; vessel less than 50 m in length	All-round light			Aground; vessel less than 50 m in length	All-round lights in a vertical line; plus white anchor light in forepart of vessel		

FIGURE 8.3

Lights and day shapes for some of the most common commercial vessels that a research vessel under 50 m in length, such as shown in Figure 3.1, might encounter. Note the construction of the lights on the mast in Figure 3.1, designed to project light horizontally through the proper angles as shown in Figure 8.1.

The rules require certain visibilities for lights. Vessels of 50 m or more in length require that the masthead light be visible 6 nm on a clear night; all other lights need to be visible 3 nm. For the ship mentioned in Chapter 3, that is, LOA less than 50 m, the masthead light must be visible 5 nm and the other lights 2 nm. Small boats, such as one less than 12 m in length, can combine the masthead and stern light into one all-round white light visible 2 nm, with sidelights visible 1 nm. Power-driven vessels less than 7 m in length and whose speed is less than 7 knots may carry a single all-round white light (sidelights being optional on this class boat). In general, a white light is required even on something partly submerged.

A three-dimensional view of a vessel less than 50 m in length that is restricted in her ability to maneuver is drawn and shown in Figure 8.4. Note the term "making way," which means she is being propelled by machinery through the water and not just drifting with the current. If the propulsion engines are stopped, the "running lights," that is, the masthead light, the sidelights, and the stern light, would be switched off. If she anchors, then the red–white–red lights remain lit and an anchor light is added. When the research vessel is now able to maneuver as required by the rules, she would show only her running lights and obey the steering and sailing rules for a vessel of her class.

If your research vessel is engaged in diving operations, she would show at night the all-around red–white–red lights described earlier and shown in Figures 8.3 and 8.4, and in addition, the International Code flag "A" (Figure 3.2) rigged to be at least 1 m high and standing stiff even without wind. The much favored red dive flag with a white diagonal stripe is not official! If the dive boat is small and cannot carry all the lights shown in Figure 8.3, she must show the red–white–red array at night and the 1-m code flag "A" by day.

Lights and shapes for other vessels such as minesweepers (three lights in a green triangle), vessels constrained by their draft (three red lights in a vertical line), pilot boats (white over red all-round lights), dredges (red–white–red plus lights to show the obstruction), tricolored masthead lantern for sailboats less than 20 m in length, and so forth, are all provided for in the 72 COLREGS. Sailors have invented memory aids to remember each designation such as *red over white fishing tonight, white over red pilot ahead.* The variety of lights and shapes may seem endless, but with time and experience they all can be learned.

Sound and light signals are the last part of the rules. A vessel of 12 m LOA or more must have a whistle; a vessel of 20 m LOA or more must also have a bell; a vessel of 100 m LOA or more must have, in addition to the whistle and bell, a gong. Small boats, those less than 12 m in length, must have some means of making an efficient sound signal. Sound and light signals warn other vessels of your presence, tell of your intentions, and signal danger. Figure 8.5 summarizes much of the detail, again focusing on a research vessel whose LOA is less than 50 m such as depicted in Figure 8.4.

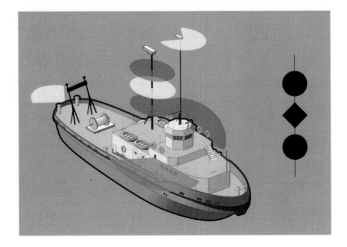

FIGURE 8.4
"Vessel restricted in her ability to maneuver—making way; vessel less than 50 m in length."
Note the flying bridge above the pilothouse.

Sound Signals

A short whistle blast is 1 second in duration, and a prolonged blast is
4–6 seconds in duration. The interval between blasts on the whistle is about
1 second. The signals may also be sent by signal light, and if so, need to be
synchronized. The heavy horizontal line in the center of Figure 8.5 separates
signals when vessels are in sight of one another with those signals in fog,
mist, snow, sandstorms, or any other condition restricting visibility.

Inland and international rules regarding signals are quite different in
meeting, overtaking, and crossing situations. The international meanings
are given in Figure 8.5, and are *announcements* of action. In inland rules, a one
or two short blast is a *proposal* of action. For example, if in inland waters your
vessel is overtaking another, a one blast signal proposes to pass on the other
vessel's starboard side. If the vessel being passed agrees, she will answer
with one blast; if she disagrees (say she sees a danger ahead) she will sound
five or more short blasts, the danger signal. The danger signal can be used
whenever one vessel is in doubt or confused about the actions of another
vessel. Cross signals are not permitted in inland waters—that is answering
a one blast signal with two blasts, and so on.

As a seagoing scientist or engineer, you may never need to know the Rules
of the Road, but if you operate a vessel, say a 12 m research vessel or a personal pleasure craft, you must know and understand them. This is a life or
death responsibility.

SIGNAL	VISUALIZE	CIRCUMSTANCE	INTERNATIONAL	INLAND
Short blast	●	In sight of one another	I am altering my course to starboard	I intend to leave you on my port side
Two short blasts	● ●	In sight of one another	I am altering my course to port	I intend to leave you on my starboard side
Three short blasts	● ● ●	In sight of one another	I am operating astern propulsion	I am operating astern propulsion
Two prolonged blasts, one short	▬ ▬ ●	Narrow channel or fairway	I intend to overtake on your starboard side	
Two prolonged blasts, two short	▬ ▬ ● ●	Narrow channel or fairway	I intend to overtake on your port side	
Prolonged, short repeat	▬ ● ▬ ●	In sight of one another	Affirmative, agree	
At least five short blasts	● ● ● ● ●	In sight of one another	Danger	Danger
One prolonged blast	▬	Obstructed vision	Announcing presence	Announcing presence
One prolonged blast	▬	Restricted visibility	Every 2 minutes	Every 2 minutes
Two prolonged blasts	▬ ▬	Restricted visibility	Not making way	Not making way
One prolonged, two short	▬ ● ●	Restricted visibility	Sailboat, fishing, towing, not under command, unable to maneuver	
One prolonged, three short	▬ ● ● ●	Restricted visibility	Last vessel of a tow if manned	
Ring bell rapidly	◄──►	Restricted visibility	At anchor	At anchor
3 strokes, ring, 3 strokes	┼┼┼◄──►┼┼┼	Restricted visibility	Aground	Aground
Four short blasts	● ● ● ●	Restricted visibility	Pilot boat on duty	Pilot boat on duty

FIGURE 8.5
Sound and light signals for vessels under 50 m in length. Note the one, two, and three short blast signals have the same meaning as Morse Code letters "E," "I," and "S"; the prolonged-short-prolonged-short is "C" (q.v. Figure 3.2).

Additional Reading

Navigation Rules, International-Inland, 1989. *U.S. Department of Homeland Security, United States Coast Guard.* Arcata, CA: Paradise Publications Inc., 221 pp.

Exercises

1. The research vessel shown in Figure 3.1 requires certain navigation lights and shapes. List the lights this class vessel requires under 72 COLREGS, their colors, their arc of visibility, and their orientation; similarly for the day shapes required. What is the relationship between the arc of visibility and the rules for avoiding collisions?

2. As the operator of a 25-foot LOA research boat, you are required to have aboard certain signals available in case of emergency. Make a list of the means of signaling an emergency, including using VHF Channel 16.

3. What is it? A nineteenth-century navigation instrument that will rattle your teeth.

9

Marlinspike Seamanship—Lines, Knots, Splices, Blocks, Tackle, Cleats, and Fairleads

Cordage, the general term for ropes and twine, probably started as vines or other natural fibers including hair and leather, and is seen in European prehistory 28,000 years ago on fired clay. Laid rope, rope made from twisted fibers, is dated to some 17,000 years ago, and was well developed by Native Americas when Europeans first came to the western hemisphere. Laid rope as an industry is recorded in Egyptian drawings about 6,000 years before the present, and is shown being used by ships as well as for landlubber purposes. It was probably used by sailors many years before *ca.* 4000 BC, and is an essential tool today on modern ships, where it is almost always called "line" (traditionally there are seven "ropes" on a naval ship: man, head, hand, foot, bell, buoy, and dip).

Rope was initially only made from natural fibers such as manila, hemp, or sisal. Synthetic fibers were added to the industry in the 1950s with the invention of nylon, polypropylene, and Kevlar. The construction can be laid, woven, braided, and so on, with natural fibers being almost always twisted into yarns, which in turn are grouped and twisted in the opposite direction into strands, which are then grouped and twisted in the opposite direction again into rope. Typical shipboard laid rope is right-hand lay of three strands—that is, the fibers are twisted clockwise into yarns, the gathered yarns counterclockwise into strands, and the gathered strands clockwise into rope. This counter twisting gives cordage its many valuable characteristics.

The marlinspike is a most useful tool when working with line. It is shaped much like the bill of a marlin or sailfish, that is, a tapered cone originally made of wood or bone. A sailor's pocketknife will have a sheepsfoot blade for cutting, and a marlinspike for separating strands. Marlinspike seamanship then is the art of customizing line aboard ship to tie up to a pier (mooring line), throw to the pier (heaving line), fasten a plankton net (towing line), prevent one from being washed overboard (safety line), or just tying down equipment in case of rough weather (tie-down line).

Shipboard Cordage

Line is cut into useful lengths from a spool. At the cut it tends to fray or unravel, and thus needs to be finished in a seamanlike manner. With synthetic line, using a shrink wrap (tube covering the bitter end that is heated

FIGURE 9.1
Sailmakers awl used to whip a natural fiber line. The awl is used to penetrate the line, create a loop, and withdraw leaving the loop. The loop is laid down along the line and the thread then is served over the loop toward the bitter end. The awl is used again to penetrate the line through the loop where the thread is cut, the awl withdrawn, and the loop pulled tight under about half of the area served. Trim off the Irish pennants to finish the whipping.

and grips the line or partially melts) is a better method than just singeing the end, which can leave rough or sharp edges. For laid rope, whipping is the preferred method, which is serving (wrapping) the line with sail twine near the end. Figure 9.1 shows the use of an awl to penetrate the strands, make a loop to tuck the twine, and serve the line to create a whipped end. Note that the amount of serving is about equal to the line's diameter, and the distance to the bitter end is also about one diameter. Also note that the whipping always serves against the lay.

Once the ends of the line are whipped (or shrink-wrapped in the case of synthetic line), it is ready for use on the research vessel. As a seagoing scientist or engineer, you will be handling instruments, and often lines are used to control the instrument from swinging in the critical space between the deck and the water. A large conductivity, temperature, depth (CTD) with rosette water sample bottles can weigh several hundred pounds, can cost $100,000 or more, and is a pendulum as the unit is being moved by the A-frame to deploy it. Passing a line through the CTD frame, one on each side, without any knots so the line can be slipped once the unit is in the water, is an important line handling task often assigned to the science party. This is a good time to be wearing a hard hat and a "stern vest" (a life jacket designed for deck work).

Knots

Knots, and there are hundreds of them, should have three characteristics: (1) easy to tie, (2) strong for their purpose, and (3) easy to untie. Perhaps the classic example, counter-example, is the square knot and the granny knot.

Both are easy to tie, but the granny knot is not strong, nor is it easy to untie (it jams). A square knot, also called a reef knot, is excellent to bend two lines of the same diameter together, and is the well-known left over right, wrap, then right over left, and wrap again (Figure 9.2). The granny knot is left over left and left over left again. A fun knot to look up is the thief's knot, which looks like a square knot, but when put under strain collapses (tradition has it that sailors would use the thief's knot to see if anyone untied their seabag, and after the nefarious deed, tied the seabag back with a square knot instead).

Line passing over a sheave in a block will run out the swallow (Figure 5.2) unless it has a stopper knot of some kind. There are several, but the two simplest are the overhand knot and the figure-eight knot (Figure 9.3). If an extra turn is taken, a double figure-eight knot is made, and it provides a larger area to prevent the line from running free through the swallow. You will recognize the overhand knot as the first phase of tying a square knot, left over right, wrap.

Perhaps the most important knot is the bowline (pronounced bo-lin; probably Middle English or Middle German in origin: bow + line). It is used on research vessels in a wide variety of ways: tying a tow line on a plankton net; a safety line around one's chest; an eye to throw over a bit; and so on. There are many ways to tie one, such as "the rabbit comes out of the hole, goes around the tree, and goes back down into the hole." The George Riser (Figure 2.1) method is a bit

FIGURE 9.2
Square knot or a reef knot, tied by taking the line end in the left hand and crossing it over the one in the right hand, then taking the line end on the right and crossing it over the end on the left.

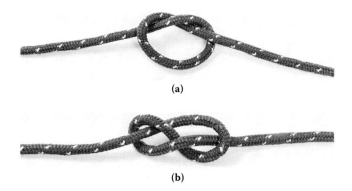

(a)

(b)

FIGURE 9.3
(a) Overhand knot and (b) figure-eight knot.

difficult to master, but it affords the advantage of being able to tie the bowline with one hand if needed. His method is to cross the bitter end over the standing part, twisting your wrist in such a way that the "rabbit" and the "hole" are made in a single motion, then wrapping the rabbit around the "tree" and back down the "hole." Once learned, the "Riser method" becomes second nature, is very fast, and results in a perfect knot each time (Figure 9.4).

If a lariat is needed, the bitter end can be led through the bight; this is called a running bowline and can be tied without needing to feed the bitter end through the bight. A bowline on a bight provides two loops and can be used to create a work harness, one loop for each leg. There are French bowlines, Spanish bowlines, Eskimo bowlines, Dutchman's bowlines, fisherman's bowlines, and fool's bowlines. Several online websites animate how to tie knots.

If bending (joining) two lines of significantly different diameter or if one is stiff as would be a hawser (heavy mooring line), the knot of choice is the sheet bend. The sheet bend was first named in English by David Steel in *The Elements and Practice of Rigging and Seamanship* in 1794, and is also known as the becket bend and the weaver's knot. A hawser will have an eye, and to attach it to a heaving line, the sailor will tie a sheet bend (Figure 9.5). The lighter line is fed up through the eye of the heavy line, wraps around both parts of the eye, and then tucks under itself as shown. If a second round turn is taken, the knot is called a double sheet bend. On sailboats the mainsheet is the line that is attached to the boom, and the jibsheet is the line attached to the clew of the jib or foresail; the term sheet bend seems not to be related to sails directly.

FIGURE 9.4
A bowline tied in synthetic line. The knot is equally useful and efficient on natural fiber line.

FIGURE 9.5
Sheet bend, used often for bending two lines of different diameter or if one is very stiff. Tends to work loose if not under strain.

A clove hitch is another useful knot that has tied many horses to a rail. In seamanship, the clove hitch is handy when tying a mooring line to a vertical post, especially if the top of the post is uncapped (i.e., not a stanchion). Figure 9.6 is a photograph of the clove hitch around a small piece of dowel. It is tied by taking a round turn that crosses over the standing part, and taking a second round turn that tucks under the first. If to a post, take two underhanded loops, cross the lower over the upper, and toss the doubled loops over the top of the post and snug up. Like the sheet bend, the clove hitch loosens if the strain is not kept taut.

The round turn and two half-hitches is another very useful knot, especially if there is a strain on a line, and the seagoing scientist or engineer wants to keep the strain. It can be tied around a post or a rail or a stanchion as shown in Figure 9.7. Pass the bitter end over the rail, and wrap it fully around 540°, that is fully around 360° and then half again. This will give enough friction

FIGURE 9.6
Clove hitch. Useful for tying a boat to a post (or a horse to a rail). Tends to become undone if not under steady strain.

FIGURE 9.7
Round turn and two half-hitches. Note that the half-hitches are made in the same direction and do not form a U shape. If in a U shape, it is called the sailor's knot, but is not as robust as the two half-hitch layout.

so that the tightness in the line will be maintained. Then over the standing part make two identical half-hitches in the same direction, and snug up the knot to the rail half-hitch by half-hitch. If when making the first half-hitch, a second turn is taken around the standing part before crossing over to make the second half-hitch, you have tied a taut-line hitch, which is handy if you want to take out the slack in the standing part and tighten up the line. Campers use the taut-line hitch for setting up a tent.

Splices

Splicing is a method of joining two pieces of line or wire rope (a short splice or a long splice), but can also be used to create a loop (an eye splice) or terminate the line (an end splice). A splice is most recognizable in three-strand line, either natural fiber or synthetic, but synthetic braided line can be made into an eye splice as well. A splice is a form of weaving the strands together to form the eye or join the two pieces or terminate without whipping. Many lines used by seagoing scientists or engineers will have eye splices with a thimble inserted to spread the load, such as would be used on a lifting bridle for a trawl-resistant acoustic Doppler current profiler (ADCP) housing. A thimble is a tear-drop shaped molded object, grooved to support the line and shape the eye. It is found on wire rope as well.

Making an eye splice in the half-inch diameter line in Figure 9.8 starts with a whipping about 6–8 inches from the bitter end. The whipping need not be a full diameter in length, but should be added before unlaying the line. It often is best to whip or tape ends of the three separate strands so that they do not unravel as the splice is being made as well. The eye is formed as seen in Figure 9.8, and the strands are laid out over one strand and under the adjacent strand, always weaving in the same direction (clockwise in this example). The usual practice is to weave over and under so that at least three rotations are completed. The eye splice is completed by trimming the strands used for the over–under tucks, and sometimes is covered with chafing gear (canvas or serving line). The eye splice is a very useful way to terminate a line, and it forms the basis for making a bridle to lift an instrument or for a tow line.

FIGURE 9.8
Eye splice in three-strand natural fiber line.

A back splice is started in the same way as the eye splice: make a short whipping about 6–8 inches from the bitter end, unlay the line down to the whipping, and whip or tape the three strands. Then a crown knot is tied, whose purpose is to reverse the direction of the strands 180°. In Figure 9.9, the crown knot is formed by laying the right-hand strand over the center strand, and passing the left-hand strand through the eye in the center strand. After the crown knot is snugged up to the whipping, the splice is woven as with the eye splice: over–under, over–under, and over–under, always rotating the line in the same direction. It is finished in the same way as the eye splice, trimming unneeded strands.

The short splice is used to fasten two lengths of line of the same diameter. It is stronger than any knot, has a smaller diameter for passing through a block, but is not of uniform thickness so may jam in the swallow if the block is not carefully chosen. The short splice is started in the same way as the eye splice or the end splice, that is, by whipping each line about 6–8 inches from the bitter end and terminating the strand's bitter end with tape or whipping cord. The two parts of the short splice are then intertwined like fingers in two hands for prayer. The three stands on the right are tied off with a bit of twine (or tape) and the over–under rotate pattern of the eye splice and the end splice is repeated first toward the left and then toward the right. When finished, there will be a total of at least six tucks and the excess stands are trimmed. A short splice is shown in Figure 9.10.

A long splice (not shown) is made by unlaying the strand from one piece of line and laying the strand from the other in the grove of the retreating strand. It starts like a short splice, but several feet of line are unlaid in the half-inch diameter example being discussed in Figures 9.7 through 9.10. There is no over–under rotate action, and the final splice is the same diameter as the original line. The long splice can also be used to splice together two wire ropes of slightly different diameter to create a tapered wire rope, which is

FIGURE 9.9
Back splice, also called an end splice, in three-strand natural fiber line.

FIGURE 9.10
Short splice in three-strand natural fiber line.

often found on deep sea moorings to save weight. The long splice is not as strong as the short splice, but it passes through a block smoothly.

Block and Tackle

For heavy lifting, say a portable laboratory van or a specialized winch, a block and tackle (pronounced "take-ell") is often needed. In elementary physics, all seagoing scientists and engineers learned about the mechanical advantage of a system of blocks, but for handy reference, see Table 9.1. The block and tackle shown in Figure 9.11 is of a twofold purchase, whose mechanical advantage in the absence of friction is 4:1, but with friction is about 2.9:1. A block and

TABLE 9.1

Block and Tackle Configurations with Their Nautical Names, Theoretical Mechanical Advantage (Left Subcolumn), and Practical Mechanical Advantage (Right Subcolumn)

Name	Rigging	Mechanical Advantage	
Single whip	Single block, single sheave	1	0.9
Gun tackle	Two single sheave blocks	2	1.7
Single luff tackle	Double sheave block, single sheave block	3	2.3
Twofold purchase	Two double sheave blocks	4	2.9
Double luff tackle	Triple sheave block, double sheave block	5	3.3
Threefold purchase	Two triple sheave blocks	6	3.8

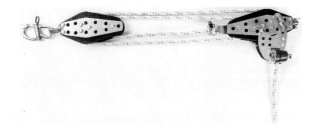

FIGURE 9.11

Twofold purchase, as might be used as the main sheet on a larger sailboat. In this particular block and tackle, the sheaves are one above the other rather than side-by-side. Note the quick release stainless steel shackle on the left for rapid reconfiguration of the attachment points. The standing part is the line from one block to another; the hauling part is the line that is pulled or released.

tackle is also useful for securing objects in case of heavy seas, or just dragging a weight across the deck. Block size (cheek length) should be about three times the circumference of the line, and the sheave diameter should be about two times the line's circumference. It is critical that the working load limit (WLL) specified by the manufacturer be known and never exceeded.

Cleats and Fairleads

Any line can be tied off on a cleat or Sampson post or bits (Figure 3.4), or other such structures meant to hold fast a tension. The general idea is to take a full 360° round turn before starting a figure-eight pattern. The friction of the round turn will keep most of the tension constant, so only a few figure-eight turns are needed. A mooring line is led through an oval-shaped structure called a chock (Figure 3.1) that is rounded to prevent chaffing.

When using the mooring line to winch the vessel close to the pier, there will be a transfer of the tension on the line to the bits. This is done by leading the line between the bits and tying it off temporarily with a stopper hitch (Figure 9.12). The stopper hitch will hold the tension for a short time, just enough to wrap the mooring line around the bits in a figure-eight pattern. The stopper hitch is then quickly released so that the bits take up all the strain. A stopper hitch is useful for other work such as transferring a plankton net towing line from a capstan or gypsy head.

A capstan is an electorally powered flanged vertical drum, about knee high from the deck to the top, used for hauling in a line such as a mooring line or a towing line. A gypsy head is a horizontal drum for the same purpose, often integrated into the anchor windlass. Both a capstan and a gypsy head drum have raised ends (I-shaped flangers) to prevent the line from slipping off. Passing the mooring line around the capstan, through the bits and chock, and overboard to a pier, often requires a heaving line that is tied to the mooring line eye with a sheet bend. The end of the heaving line will have a monkey's fist for mass, and is a fun knot to learn to tie (but be careful as it is considered a weapon in some places).

FIGURE 9.12
Stopper hitch used at bits between the stopper and the mooring line to transfer strain from a capstan or gypsy head to the deck structure. Bits (not shown) are to the left in the figure; strain is to the right. A slightly modified version is useful in camping for tensioning a line on a tent.

Additional Reading

Noel, J. V., Jr. 1988. *Knight's Modern Seamanship*, 18th Edition. New York, NY: John
    Wiley and Sons, 800 pp.

Exercises

1. The seagoing scientist and engineer should have some notion of the working
   load limit (WLL) of various types of line, natural or synthetic (see also the
   replaced term: SWL). Define WLL, and make an Internet search for such
   information. In particular look at the Cordage Institute website (http://www.
   ropecord.com/new/index.php).
2. Learning to tie knots and make splices is best done in a hands-on setting. Learn
   to make the following: square knot, bowline, sheet bend, clove hitch, round
   turn and two half-hitches, stopper hitch, eye splice, and short splice. Find a
   website to help you learn.
3. "What is it?" Think bird . . .

10

Trailerable Boats—Hubs and Hitches, Trailering, Boat Ramps, Launching, Recovery, and Anchoring

A small boat trailer is technically a cart, a two-wheeled vehicle like a chariot. The history of moving heavy objects goes back to at least *ca.* 7000 BC when sledges were in use, but the dog travois and the horse travois probably date back to the last ice age. In 3000 BC, wagons with two axles and four solid-wood wheels are known to have been used in present day Switzerland. By 2000 BC, wheeled transports were in use from Europe to Persia and were considered so important that they are found in ancient tombs for transport in the afterlife. The first-wheeled vehicle to carry a boat is not known, but it is hard to imagine that someone in ancient times did not see the obvious use.

Modern boat trailers are made with metal frames, either galvanized steel or aluminum, and are fitted with a spare tire, towing lights, and a license plate bracket. Usually, a cross frame box beam is connected to the frame by leaf springs, which are held in place with U-bolts. The solid wheel hub axles, which are about 10″ in length, are welded into the box-beam. The hubs turn on bearings fitted to the axle, and must be maintained regularly by packing them with waterproof grease. Nothing can be quite so disconcerting when towing than for a bearing to burn out or freeze at 55 mph; salt water has a nasty way of infiltrating seals and hubs. Maintaining the leaf springs too is essential as they are, of necessity, ferrous materials. Figure 10.1 shows an assembled wheel hub, and Figure 10.2 shows the disassembled unit.

Trailer hitches are familiar to most everybody, but it is essential to know that trailer balls in the United States for recreational use come in different diameters: 1 7/8″, 2″, 2 5/16″, and occasionally 3″. The ball size required is stamped on the top of the trailer coupler, which is the mechanism that secures the ball on the towing vehicle to the trailer. Once the coupler settles onto the ball, a release leaver secures the ball-coupler hitch. The release lever should be pinned shut, usually with a padlock, and the safety chains attached. Safety chains are attached by crisscrossing the left chain under the right chain, and placing the hooks facing the trailer into the eyes of the receiver hitch on the towing vehicle. Lastly, the electrical connection needs to be made, and all lights checked for proper function—brake lights and turn signal lights.

FIGURE 10.1
Wheel hub of a small trailer used in towing boats. From left to right: hub cap, cast and machined hub, lug nuts, bronze lugs, and inner part of the hub axle that fits into the box-beam where it is welded in place. The tire and wheel are mounted from the left and held in place with four lug nuts.

FIGURE 10.2
Wheel hub assembly; parts from left to right are axle, seal, inner bearing race, machined hub with lug nuts, outer bearing race, compression washer, lock nut, cotter pin, and hub cap. The four lug nuts hold the wheel and tire assembly in place. The bearing race must be completely filled with grease to lubricate the bearing rollers, and prevent water from washing out the lubricant.

Trailering, Launching, and Recovery

Driving with a trailer requires substantially more attention to the road and road conditions than just moseying down life's highway. The towing vehicle should weigh at least as much as the boat/trailer being towed, and often a factory-fitted towing package is necessary for the towing vehicle. The "rig," the combination of tow vehicle and trailer, requires wider turns, accelerates slower, and takes longer to stop. Fuel consumption increases, engine cooling is less efficient, and crosswinds are a major concern. Passing vehicles such as an 18-wheeler can break the airstream, and if on slick pavement or ice, may cause your rig to jackknife and/or spin out of control. If the boat being towed is heavy, it is best to use the trailer brakes to slow down so as to drag the towing vehicle to a stop. Trailers for a small runabout will not have brakes, and thus extra caution is required bringing the rig to a stop. The two-second rule of space between your rig and the vehicle ahead is not sufficient; leave at least 5 seconds of space-time.

Trailer tires are vulnerable to losing air because their tread wear is small yet they tend to dry out and form cracks. If the tire goes flat, pull off the road sufficiently that passing vehicles will not endanger you. Set the jack under the trailer's axle, block the opposite wheel(s), and begin to lift the trailer. Loosen the lug nuts before jacking up the wheel, and set a jack-stand under the frame. Remove the lug nuts, remove the wheel, and place the spare on the hub so that the chamfer on the lug nuts fits into the wheel properly; hand tighten the lug nuts before removing the jack-stand and lowering the wheel. Once the spare tire is on the ground, tighten the lug nuts in a star pattern, remove the jack and the wheel blocks, and check the rig before resuming the trip.

Ramp courtesy means having the boat ready to go before launching her. Items for safe and productive passage are loaded into the boat after arriving at the launching marina, but before approaching the ramp. Table 10.1 is a checklist of items every small craft should carry; other items may be added such as a cooler of food and drinks, prescription medicines, a paddle, and a seaman's handbook. Be sure that the boat's running lights are all work-ing; you could be returning after dark due to unforeseen circumstances. The weather should have been checked the day before, and a cruise plan filed with the office and/or a family member. Weather conditions can change drastically from a bright sunny morning to massive cumulonimbus systems by afternoon; the prudent mariner must be prepared.

Backing a trailer is a skill acquired by experience. Backing straight is almost impossible without using the steering wheel to make small changes in direction. Find an empty parking lot or lawn, and practice backing.

TABLE 10.1

Small Boat Launching Checklist (items should be loaded before approaching the ramp so as not to cause other users to wait unnecessarily)

Item	
Anchor and rode	Batteries; charged, engine
Bilge pump	Cellphone and/or very high frequency (VHF) radio
Chart	Compass
Dock lines; spare line	Fenders
Fire extinguisher	First aid kit
Flares (at least three)	Flashlight; spare batteries
Foul weather gear	Fuel/oil + spare can
GPS receiver; spare batteries	Keys with float
Knife with marlinspike	Life preservers (one for each person)
Life ring (throwable)	Plug (and a spare)
Sunscreen	Tool kit
Water, potable	Whistle or CO_2 horn

Remember that the direction of the bottom of the steering wheel is the direction that the trailer will turn: to back the trailer to the left, the steering wheel is turned to the right (bottom to the left), and *vice versa*. Make small adjustments, foot on the brake, using slow deliberate movements. If the trailer is not going where it needs to go: stop, go forward slowly, straighten out the rig, and try again. If you have a partner, ask him or her to go where he/she can be seen in your side-view mirrors, and help direct the trailer to the space where it is wanted.

Backing a boat and trailer down a boat ramp is different from doing so in a parking lot because the ramp may be slippery, and erosion at the end of the concrete apron may have left a shallow ditch. Having a four-wheel drive towing vehicle is a plus. The best approach is to straighten out the rig before backing down the ramp if possible. Align the boat with the pier, leaving a foot or so of space between the boat and the pilings; have fenders at the ready. Undo the transom straps that have held the boat to the trailer (two are recommended, one about one-fourth boat length forward of the stern and another one-fourth of a boat length aft of the bow), and put in the drain plug. Slowly back down until the boat begins to float, tie a bowline and a stern line to the pier, release the trailer winch, undo the winch cable, and pull the trailer away from the boat. Ramps can be crowded, so it is prudent to be as quick as possible, but not reckless.

Recovery is pretty much the reverse of launching. Back the trailer down to the same depth where the boat began to float, set the parking brake and put the transmission in neutral. Bring the boat to the trailer centered, attach the trailer winch cable, and winch the boat into the cradle. Once the boat is two-blocked to the winch: foot on the brake, put the transmission into drive, releasing the parking brake as the towing vehicle slowly moves up the ramp to the level area or parking zone if the ramp is crowded. Strap the boat down to the trailer, remove the boat's drain plug, check to see that all electrical connections are made, remove all items that might fly off the boat at highway speeds, and prepare for the trip home. If possible, a thorough wash down and engine flush should be made before beginning to drive; this will minimize the risk of transporting invasive species from one waterbody to another.

Anchors

September 12, 1960. The USC&GS Ship *Explorer* was anchored in the lee of Cape Cod as Hurricane Donna passed over the Bay. The *Explorer*, a 220′ hydrographic survey vessel, had been updating nautical charts on Nantucket Shoals when the storm began to approach; Cape Cod would provide a degree of shelter from the 100 knot wind gusts that had devastated the Outer Banks of North Carolina. Two standard stockless anchors were deployed in a "Y" with the *Explorer* at the apex of the "Y," engines were slow ahead, and the *Explorer* was dragging anchor! She sustained only minor damage, but two 1000 pound stockless naval anchors and several shots of chain (1 shot = 15 fathoms) were not enough to keep her from dragging anchor and going backwards.

Anchoring is the art of staying put. Ground tackle is the general term that includes the anchor, anchor chain, and on smaller vessels, the rode (shackles, chain, and line). A vessel such as the *Explorer* uses an all-chain ground tackle, whereas a smaller vessel, such as the trailered boat discussed earlier, usually carries an anchor and a rode made of a short length of chain and (usually) a braided nylon anchor line, because nylon stretches 10%–15%, and the stretching acts as a shock absorber. On a larger vessel such as the *Explorer*, the anchor is housed in a hawse pipe, but on a smaller vessel such as the one in Figure 3.4, it is secured in an anchor chock.

Anchors come in a wide variety of designs. In the nineteenth century and earlier, a stock anchor was used (see Figure 14.6), which had flukes for digging into the bottom, a crown rigidly connecting the shank to the flukes, a stock which is fitted through the shank at right angles to the flukes, and a ring which is connected to the chain by a shackle. Larger ships use a "naval" anchor, which is stockless in design, but with flukes hinged at the crown. There also are Danforth anchors, plow anchors, Bruce anchors, folding anchors, and mushroom anchors; the list goes on. Each has its specialty for holding in sand, shell, mud, silt, and so on, but for smaller boats, the Danforth anchor is very serviceable. Figure 10.3 is a US Navy lightweight (LWT) anchor as might be used on a research vessel such as pictured in Figure 3.1.

The rode for the anchor in Figure 10.3 should have about 600' of half-inch stud-link chain connected to the anchor shackle. The chain lays on the bottom and acts to make the pull on the anchor horizontal. This forces the flukes to dig in and hold. For a smaller boat, such as say a 21-foot trailerable skiff envisioned in the earlier discussion on trailering, a Danforth anchor should weigh about 8 pounds, and have a rode of 12 feet of 3/16 inch chain attached to 150 feet of 3/8 inch braided nylon. Carrying a second smaller spare anchor is always an act of good seamanship.

Connecting the braided nylon rode to the anchor chain is best done with an eye spice using a nylon thimble (Chapter 9). If the line is connected directly to the chain, an anchor hitch is used as shown in Figure 10.4.

Before deciding to anchor, consult the latest nautical chart. Anchoring may be prohibited in certain areas, and special anchorages are often shown on the chart. Choose an area with a sandy bottom if possible; sand is shown on the chart by an "S"; mud, silt, and shell are poorer holding bottoms. The scope of rode should always be at least three times the water depth; 5:1 and 7:1 are better, especially if the bottom is silt or mud. Be sure there is sufficient sea room for the ship to swing a full 360° as wind and tidal currents back or veer

FIGURE 10.3

US Navy 200 pound lightweight anchor. For scale, the shank is about 36″. From top to bottom, the parts are anchor shackle, shank, flukes, crown, stock, and tripping ring. The stock acts to lay the flukes down and dig into the bottom. Anchors of this type are commonly called Danforth anchors. "Danforth-style" anchors are particularly good for sandy bottoms.

FIGURE 10.4

Anchor hitch, also known as a fisherman's bend, through the crown of a shackle. From right to left, the parts of a shackle are lugs, the jaw (opening), the bolt, the clear (circular area), and the crown. The knot is tied by taking two round turns around the shackle crown, coming through the two round turns, and finishing it with a half-hitch. The bitter end often is sewn with an awl onto the hauling part for additional security.

with time. Allow as much room to swing on the anchors as rode is let out; banging into a neighboring ship is not generally appreciated too much.

Anchoring

When anchoring, bring the ship upwind slowly to the point where the anchor is to be set. Lower, do not drop, the anchor to the bottom, and pay out the rode as the ship drifts sternward or slowly backs down. Once the line paid out is the length of the scope desired, tie off the line on a Sampson post, or if using all-chain ground tackle, secure the anchor windlass. Watch the chain or line as appropriate, to see if the ship stays in place. The chain or line should not be tight and leading directly to the anchor; a catenary is desired. Once sure that the anchor is holding, take visual bearings or GPS coordinates or RADAR fixes (Chapter 7), and write them down in the log. Checking the position should be a regular duty of the person on watch.

Weighing anchor, the act of bringing it back aboard, should be done using the engines to slowly move toward the spot where the anchor is deployed. Haul in on the anchor windlass or gypsy head as the ship moves forward, being certain that the rode does not get tangled in the propellers. Once the ship is directly over the anchor, the windlass or gypsy head should be able to break it free and hall it in. If the anchor does not readily break free, slowly drive over the spot or go in a tight circle and pull from upwind or up-current. On a rocky bottom, the anchor may well get stuck, and if no divers are abroad, it may have to be abandoned.

If your vessel is disabled in deep water, and unable to "drop the hook," a sea anchor may be needed. A sea anchor will keep the bow into the wind and waves, but will not stop the vessel from drifting. Tie a mooring line around some flotsam (say an ice chest or a plankton net or a sea anchor if one is aboard), toss it overboard at the bow, feed out the line, and make the line fast to a cleat through the bow chock. The drag on the sea anchor will keep the bow pointing into the weather for the most part, and reduce the risk of capsizing. It will also slow the rate at which your boat is drifting and increase the likelihood of being rescued by decreasing the search area.

Additional Reading

Eaton, J. 2013. *Chapman Piloting & Seamanship*, 67th Edition, New York, NY: Hearst Communications, 920 pp.

Exercises

1. Draw a sketch of a boat trailer, identify the major parts, detail the procedures for hitching and unhitching, greasing the axle bearing, and changing a tire. List the tools needed.

2. John Elliott Pillsbury (1846–1919) was a US Naval Officer and oceanographer, who, among other things, was able to anchor his ship in the Gulf Stream (1888–1889) and take subsurface measurements of the current velocity. Research Pillsbury's anchoring technique and give a short report. How would you anchor a 150 foot LOA research vessel today in the Straits of Florida between Miami and Bimini? In your answer, detail the type of anchor, rode (chain, wire rope, shackles, etc.), and scope, in order to stay on station in a three knot surface current with water depth of 400 fathoms. Be specific regarding sizes, materials, and equipment; include a drawing of the rode.

3. "What is it?"

11

Handling Equipment—Superstructure, Deck Machinery, Wire Ropes, Clothing, and Commands

Posidonius, *ca.* 100 BC, is reputed to have made a sounding of 1000 fathoms, and never reached the bottom. Whether he did or not may be lost to history, but the notion of lowering a weight, probably a stone, on a line made of natural fiber, and trying to determine water depth is well in keeping with his character as a stoic philosopher. Capstans were known in Roman times, so might Posidonius have used such a device to lower or haul in the 6000 feet of line needed? All with manual labor? Today, the electrically powered capstan remains a common piece of deck machinery, and perhaps is the forerunner of the oceanographic winch. Most certainly the HMS Challenger had dedicated winches, one of which was used to discover the Challenger Deep in the Marianas Trench, where a sounding of 4475 fathoms was reported (modern day estimates are approximately 5964 fathoms [35,783 $\pm$ 27 ft or 10,907 $\pm$ 8 m]).

The main deck on a ship is that deck up to which the watertight bulkheads extend; all above is superstructure. In Figure 3.4, the main deck could be also be called "B deck," with "A deck" being the deck below where the machinery spaces, tank tops, and some of the staterooms are located (see also Figure 3.5). The superstructure deck is "C deck" on the 78 ft length overall (LOA) coastal research vessel pictured in Chapter 3, and it houses the pilothouse, bunks for the crew, life rafts, and some deck machinery. On larger ships there may be many decks in the superstructure up to and including the flying bridge above the bridge (a proper "flying bridge" would have a safety rail). The term "bridge" seems to be used more often on larger ships and on naval ships; "wheelhouse" and "pilothouse" are interchangeable terms and are used on smaller ships. If a ship is large enough for the term "bridge" to be used, it most likely will have open "bridge wings," one on each side, for maneuvering in close quarters or for the lookout.

Deck machinery used in oceanography and ocean engineering includes winches, cranes, A-frames, davits, capstans, and gypsy heads. The most used is the oceanographic winch. Some winches have more than one spool of wire, typically a multiconductor wire with slip rings for electrical transmission, and a hydrographic wire for raising and lowering instruments such as water sample bottles, plankton nets, grabs and corers, dredges, current meters, and so on. Specialized instruments such as a magnetometer will most likely have their own winch and cable; mechanical bathythermographs (BTs) are rarely used today, but they did have a winch and cable prior to the invention of the expendable bathythermograph (XBT). For heavier lifting such as a submersible (Figure 4.3), a specialized A-frame and winch system is required.

A portable conductivity, temperature, depth (CTD) winch is shown in Figure 11.1. This particular winch runs on 110 V AC; other and larger winches could be 220 V and might be DC. The winch in Figure 11.1 is designed for a small CTD, and cannot be used for towing or lifting as the risk to cable damage is too great. Many CTD operations are over the stern using the A-frame, but in some cases they may be over the side usually somewhere amidships near the wet lab.

FIGURE 11.1
Portable conductivity, temperature, depth winch with base designed for standard University-National Oceanographic Laboratory System (UNOLS) 2' by 2' deck bolt pattern (see Figure 3.4). The winch is 2 ½' tall and 2' wide, and weighs 200 lbs. The wire drum in the foreground holds 1500 feet of multiconductor cable and is led through an automatic level-wind, which in turn is a fairlead to a block on the A-frame.

To insure smooth spooling of the cable, the fairlead distance to the meter wheel on the A-frame from the winch needs to be at least as great as 15 times the width of the drum. This will be accomplished if the maximum angle from the drum to the meter wheel, called the "fleet angle", is 1.5°–2°. For the winch in Figure 11.1, with a drum width of 1.5 feet, the distance should be at least fairlead = 15 × 1.5′ = 22.5 ft. That is, the arctan of 0.75/22.5 = 1.9°.

Some larger research vessels will have many miles of wire rope for deep sea dredging, coring, and anchoring. The winch on such a system will not consist of a take-up drum such as shown in Figure 11.1, but rather a separate traction winch for hauling in and lowering, with a take-up drum that only keeps tension between the traction winch and the drum. The traction winch is usually two multigrove sheaves set one behind the other; the wire wraps around the sheaves about six times in separate groves, and then is led to the take-up drum for storage. These systems can hold as much as 10,000 m (6 miles) of wire, and usually are located below-decks, although there is a University-National Oceanographic Laboratory System winch facility from which they can be borrowed.

Wire Rope

The wire rope shown in Figure 11.1 is a multiconductor cable for a CTD. For other deep sea research such as dredging or coring, plain wire rope is used. In order to save weight, the wire rope may be tapered, that is, a length of half-inch wire is spliced into the next size, say 9/16 inch, and the length of 9/16 inch is spliced into a length of 5/8 inch, and so on. The safe working load of the tapered wire rope is given by the diameter of the smallest wire. The long splice is used so that the wire can smoothly spool over the traction winch and/or the block and tackle being used.

While unbraided fiber line is usually right-twisted threads making left-twisted strands making right-twisted line, wire rope is a bit different. Wire rope might be listed as 6 × 19 meaning it is constructed of six strands with 19 wires in each strand. Galvanized 6 × 19 wire rope is very strong, and is often used for standing rigging and topping lifts where flexibility is not at a premium. It can be "right regular lay," that is similar to most fiber line, or "left regular lay" where the twists are opposite of right regular lay. "Lang lay" wire rope, left or right, is where the wires and strands are twisted in the same direction, and "reverse lay" wire rope is where the wires in one strand are twisted to the right and those in the juxtaposed strand are twisted to the left and all strands are twisted to the right.

A more flexible wire rope than 6 × 19 is 6 × 37, but while they are both of the same diameter, the individual wires in the 6 × 37 are much smaller than in 6 × 19, and more subject to corrosion and wear. For permanent standing rigging such as shrouds and stays on a sailboat, 6 × 7 wire rope is used, but it is not as flexible as the other two just discussed. Wire rope will often have a core of nonmetallic materials, or the core can be of the same strand configuration in which case, for example, 6 × 19 would be 7 × 19 construction. The combinations of wire rope construction are virtually limitless with the three six-strand examples discussed being most often used aboard ship (Figure 11.2).

Wire rope can be terminated with an eye splice just as with fiber, but more often it is accomplished with wire rope clips (sometimes called wire rope clamps) or a pressed ferrule. Two examples are shown in Figure 11.3, with the wire rope clip method on the left and the ferrule method on the right; both eye splices have a thimble to better spread the load. Note that when using wire rope clips, the U-bolt is over the bitter end of the rope and the saddle over the standing part. The clips should be spaced apart about six wire rope diameters, and must be checked periodically as vibrations can loosen the nuts. Wire rope clips are often considered a temporary eye-splice method, whereas pressed ferrules are more permanent. In either case, the clips and ferrule must be matched to the wire rope diameter, and especially the ferrule area inspected for corrosion and potential failure.

Wire ropes are prone to having kinks. A kink can occur from mishandling or an unexpected surge of the ship, and is a permanent weakness in the rope. If the kink is not so severe that the wires or strands are unbroken, it may

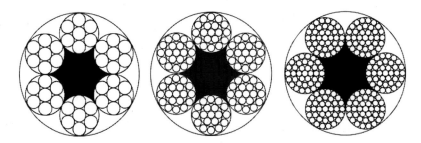

FIGURE 11.2
Wire rope construction, left to right: 6 × 7, 6 × 19, and 6 × 37, all three with a nonmetallic core. The circle shows the proper method of measuring the wire rope diameter and is not part of the construction.

FIGURE 11.3
Wire rope eye splice using wire rope clips (left) and using a pressed ferrule (right). In both cases a thimble is inserted to strengthen the eye. The U-bolt, saddle (sometimes called the roddle), and nuts are shown to illustrate the parts at the left.

be possible to straighten it out, but never by pulling on the ends on either side of the kink. Pushing the rope on either side of the kink to open a bight may save the day, after which the rope may be reshaped with a wooden mallet. If there is any doubt, cut out the kink and rejoin with a long splice. If a kink occurs in a multiconductor cable such as on the winch in Figure 11.1, the cable most likely will need to be cut, shortened, and a new watertight termination made. Rarely can such a termination be made at sea, so it is extremely important to pay continuous attention to the operation of the winch and wire.

Wire spooling on a winch drum should be smooth without overriding or cross-winding. This requires that the level-wind be matched to the wire and to the drum. Sheaves also must match the diameter of the wire rope and not have any broken flanges. When handling a CTD for example, be sure that when the instrument is on deck and a bit of slack is in the wire, that the wire does not cross-wind. That is, when taking up the slack while preparing to lift the CTD off the deck, keep the wire on the drum evenly spooled. Wear gloves!

If an instrument is attached to a wire rope eye with a shackle, mousing the shackle is a mandatory safety precaution. The screw pin of a shackle is forged with an eye for this purpose. To mouse a screw pin shackle, a piece of wire is passed around the shackle and led through the pin eye in a figure-eight pattern. Usually, three figure eights are sufficient followed by twisting the mousing wire and trimming off any excess. A moused shackle, shown in Figure 11.4, will not vibrate loose. A round pin shackle is secured by a cotter pin and does not require mousing.

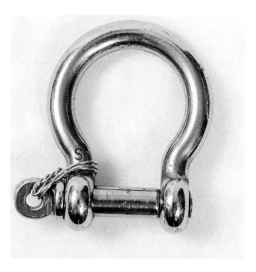

FIGURE 11.4
A screw pin shackle moused with wire to prevent the pin from unscrewing due to vibration.

Clothing

Do not bring any clothing aboard a research ship that cannot get wet. Or blown overboard. Or ripped. Or stained with grease. Or... Going to sea is going into a work environment that can be hazardous and physically challenging. Proper clothing for the weather is a must!

Every seagoing scientist or engineer needs foul weather gear, which should include a waterproof jacket with a hood, matching waterproof pants, and rubber boots. In tropical climates, getting sneakers wet is fine, but not in an arctic environment! Decks and oceanographic bucket stations are often awash in salt water. Decks will be painted with a nonskid surface, but seawater rushing out to a freezing port can be forceful enough to make you lose your footing; it can splash up the inside of foul weather pants too. Many marine supply stores have foul-weather gear selections, and be sure to put your name on the labels with waterproof ink.

If your cruise is handling heavy equipment, you may be required to have a hard-hat, work gloves, and/or steel-toed shoes. It is the chief scientist's responsibility to contact the ship and discuss work-clothes requirements. Then, of course, be sure that the scientific party is informed in a timely manner. Some ships will have extra hard-hats and steel-toed slippers into which a shoe fits. Flip-flops and sandals never are appropriate footwear on deck. Safety is the prime objective, so footwear should include closed-toed shoes such as sneakers or better yet, deck shoes.

Mess arrangements vary from ship to ship. On the small research vessel discussed in Chapters 3 and 5 (see Figure 5.1) there is one mess shared by all, and it probably is fairly informal. Many UNOLS ships also have a common mess for all aboard, but some larger ships, particularly if staffed by commissioned officers, may have separate messes. The officer's wardroom is a more formal environment, where the officers are expected to be in proper uniform. If members of the scientific party eat in a wardroom setting, t-shirts, tank-tops, cut-offs, and flip-flops are not in keeping with the decorum. Having slacks and polo shirts in your seabag may avoid an embarrassment.

And hats are *always* removed when entering the mess deck.

Larger ships will have laundry facilities, but not one like the 78 footer in Chapter 3. Plan accordingly to have sufficient clean clothes. Ships are rather cloistered environments, and smells seem amplified. Showers are communal facilities; leave it as you would like to find it. Wear flip-flops to minimize spreading athlete's foot, and take "sea showers" to save water—that is, wet down, shut off the water while soaping up, rinse quickly, and do not linger. Hang your wet towel where it will dry; mold and mildew are constant companions on a ship.

Commands

Voice communication requires using proper nautical terms. There will be moments, however, when only hand signals will do, say for example, deploying and retrieving a CTD over the stern A-frame in windy conditions. If, as in Figure 3.4, the winch operator is on the superstructure deck, he/she cannot see the CTD in the water. Hand signals such as illustrated in Figure 11.5 are the essential common language of safely lowering and raising a $100,000 instrument when voices are not the best vector of communication.

Consider the following scenario: A CTD cast in overcast marginal weather conditions at 0300 in the Gulf Stream on March 15th—the Ides of March. The wind is against the current so seas are choppy with a 6-sec period and significant wave height of 3 m. The deck officer on watch turns on the red-white-red survey lights (Figure 8.3), brings the bow into the sea, and

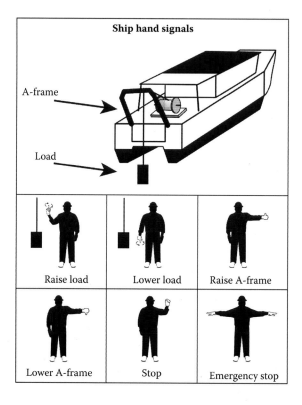

FIGURE 11.5
Hand signals between deck watch and the winch operator used to signal raising and lowering the winch and the in-and-out of the A-frame. Signals are based on crane operators, and are applicable to both A-frames and deck cranes.

slows to almost dead slow; this will keep the ship pointed into the weather and almost at zero speed over the ground (SOG) according to the Global Positioning System. The science team on watch has prepared the CTD. The Niskin bottles in the rosette are all cocked; no time is going to be wasted on this, the last cast of the current's transect. Everyone is tired, but ready to get the last cast and "head for the barn"—going home!

It begins to rain.

A crew member is at the winch on C deck, and two in the science party, in foul weather gear on the B deck fantail, stand by the CTD under the A-frame, which is in its forward-most position. Using hand signals, the CTD is slowly lifted off the deck and the A-frame is simultaneously moved aft so that the CTD will clear the stern. Carefully, the winch wire is let out as the A-Frame arcs up, and the CTD slips clear of the transom. As the CTD moves sternward, the winch wire is now let out so that the finger-grip does not jam into the meter wheel. All this "ballet" is done by sight and hand signals, but no voices are heard above the howling wind. Quickly the CTD is lowered into the water and the winch is stopped so that the surface calibration is made and the instrument does not swing like a pendulum. The mate on watch in the pilothouse watches the wire angle and by hand signals knows to urge the ship forward so the wire does not get fouled in the propellers.

The cast is ready. The chief scientist wants a profile to 2000 m depth, but the officer on watch notices that the water depth is 2200 m and slowly decreasing. As the CTD is lowered at 100 m per minute, everyone is watching the water depth—in the pilothouse, and in the dry lab where a repeater of the fathometer is displaying the data. Tension is running high as the financial equivalent of three BMW 335i motor vehicles are suspended from a 1/8 inch cable almost 1¼ miles away. The officer on watch calls the captain, the captain calls the chief scientist; a decision is needed. A team decision is needed.

Nineteen minutes into the cast, the chief scientist orders the winch stopped and the CTD returned to the surface with a few hundred meters of safety. The deck watch is searching for the CTD in the floodlight beam as it nears the surface. She waves to the winch operator and points two fingers toward her eyes and to be sure shouts "in sight." The winch is slowed almost to a dead stop as the instrument slowly approaches the surface.

Suddenly she raises her fist to signal stop the winch. Then uncharacteristically waves the deck watch to the fantail. There in the glow of the A-frame's floodlight is a huge mola mola. Everyone stares into the deep at this 12 ft "angelfish" is curiously and lazily circling the CTD. Then as suddenly as the mola mola appeared, it is gone, and a magic moment between man and fish is over. Now all attention is focused on getting the CTD safely on deck, and the data stored in the computer. Hand signals raise the CTD, rotate the A-frame, and lower the CTD to the deck. The pilothouse is signaled and the ship slowly turns to start the run to safe harbor and the end of the cruise. The scientific party on watch begin to drain the Niskin bottles to collect specialized samples, which are to be frozen for analysis ashore.

Additional Reading

Blair, C. H. 1977. *Seamanship: A Handbook for Oceanographers*. Baltimore, MD: Cornell Maritime Press, 227 pp.

Exercises

1. Knowing that the fleet angle must be less than 2°, what is the maximum drum width of the centerline winch shown on the vessel in Figure 3.4?
2. Wire ropes are used extensively on research vessels. What is the diameter of a typical hydrographic winch wire ("hydro wire")? Make a short list of cable diameters up to 1" with safe loads in both British engineering units and metric units. Convert safe force to safe mass.
3. Named for the famed organizer of the *RV Fram* expedition: What is it?

12

Oceanographic Stations— Preparation, Time, Position, Weather, Personnel, and Ancillary Data

Oceanographic stations as we know them today most likely can be credited to the work of Charles Wyville Thompson and the voyage of *HMS Challenger* (1872–1876). Certainly the explorations of Royal Navy Captain James Cook on *HM Bark Endeavor* and Sir Charles Robert Darwin on *HMS Beagle*, both later to become Fellows of the Royal Society (FRS), greatly expanded knowledge of the seashore, but not of the deep sea per se. The "Exploring Expedition" of US Navy Lieutenant Charles Wilkes (1838–1842) too was mainly carto-graphic in nature; it added to marine biology and whaling knowledge, but not to the sea floor or the water column. For that quantum leap in oceanogra-phy, the voyage of *HMS Challenger* (Figure 12.1) deserves special recognition.

In the 62,000 nm voyage of *HMS Challenger*, Charles Wyville Thompson took 354 oceanographic stations, coming (pardon the pun) almost full circle back to the United Kingdom. In addition, 504 water depth soundings were taken, some at the oceanographic stations, others separately (Figure 12.1). A standard *Challenger* oceanographic station included:

- Animal and plant life at various depths
- Seafloor samples
- Surface water variables (current, density, salinity, etc.) and some-times at depth
- Temperature at various depths
- Water samples (for later analysis)
- Water depth
- Weather

To make all these observations, over 12 miles of piano wire and 144 miles of hemp line were aboard the ship. Many unique instruments were used such as a "sounding apparatus" (200 kg weight) for water depth deter-mination, dredges for sea floor sampling, trawls for capturing plant and animal life, and the first use of Negretti and Zambra reversing thermom-eters for measuring temperature at depth. Much of the data obtained at

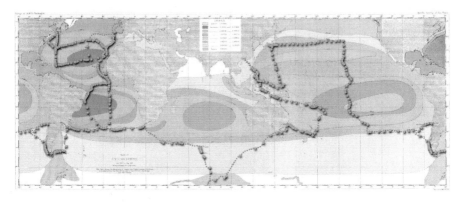

FIGURE 12.1

Track of *HMS Challenger* (1872–1876) overlaid on map of sea surface specific gravity (g·cm³). Red symbols are the ship's track; blue dots are locations of the 504 deep sea soundings. Base map from the Special Collections of the University of Amsterdam. (Sounding locations from Appendix II of *Narrative of the Cruise of H.M.S. Challenger*, T.H. Tizzard et al., 1885.)

a *Challenger* station is standard practice on modern voyages of marine discovery.

In 1955, the US Navy Hydrographic Office published the *Instruction Manual for Oceanographic Observations*. The second and third editions were also published, the last in 1968 under the title *Instruction Manual for Obtaining Oceanographic Data*, by the US Naval Oceanographic Office as Pub. No. 607. The third edition had many chapters, and each one was dedicated to (now) classical instruments such as the mechanical bathythermograph, Nansen bottles and reversing thermometers, the Phleger core for seafloor samples, the Kullenberg piston corer for sub-bottom sediment sampling, the Edgerton, Germeshausen, and Grier (EG&G) underwater camera for bottom photography, the Clarke-Bumpus quantitative plankton sampling net, Forel water color scale and Secchi disk for water transparency and chlorophyll concentration, and so on. Pub. No. 607 is a virtual history of early and mid-twentieth century ocean science and instrumentation.

To illustrate the organization and detail of an oceanographic station, Oceanographic Log Sheet-A, from Pub. No. 607, is shown in Figure 12.2. Log Sheet-A is typical of a "Nansen cast" station, that is, using Nansen bottles with reversing thermometers on a hydrographic wire for obtaining water samples and temperatures at depth. Note the detailed ancillary information on the log sheet that includes date, time, position (latitude and longitude), the ship's name (United States Ship [USS] *San Pablo* in this figure), observer's name, meter wheel (Figure 5.2) reading, Nansen bottle number, reversing thermometer serial number, salinity sample bottle number, and a complete surface weather report. Such information is critical in documenting any observation at sea, and would be titled "metadata" today.

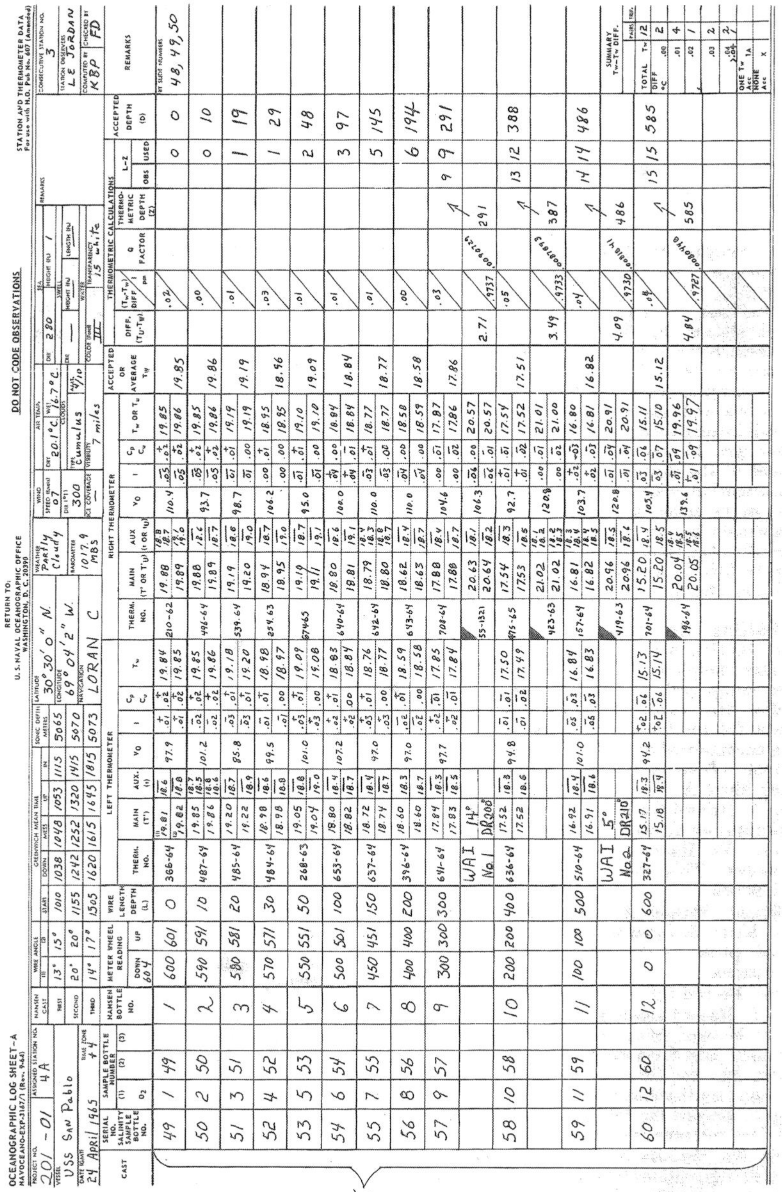

OCEANOGRAPHIC LOG SHEET-A
NAVOCEANO-EXP-3167/1 (Rev. 9-64)

RETURN TO: U.S. NAVAL OCEANOGRAPHIC OFFICE WASHINGTON, D.C. 20390

DO NOT CODE OBSERVATIONS

STATION AND THERMOMETER DATA For use with H.O. Pub No. 607 (Amended)

Project No. 201-01; Assigned station no. 4A; Consecutive station no. 3; Vessel USS San Pablo; Date (GMT) 24 April 1965; Time zone +4; Latitude 30°30'0" N; Longitude 69°04'2" W; Navigation LORAN C; Sonic depth (meters) 5065, 5070, 5073; Wind speed 07, dir 300; Weather Partly cloudy; Barometer 1017.9 MBS; Air temp. dry 20.1°C, wet 16.7°C; Clouds Cumulus 4/10; Visibility 7 miles; Sea dir 280, height 1; Water color III, transparency 15 white; Station observers LE Jordan; Computed by KBP; Checked by FD

Serial No.	Sample bottle (1)	(2)	Nansen bottle no.	Meter wheel down	Meter wheel up	Wire depth (L)	Left therm. no.	Left main (T')	Left T_w	Right therm. no.	Right main	Right T_w	Accepted average T_w	Diff. (T_u-T_w)	Thermometric depth (Z)	L-Z obs	L-Z used	Accepted depth (D)
49	1	49	1	600	601	0	366-64	19.81 / 19.82	19.84 / 19.85	210-62	19.88 / 19.89	19.85 / 19.86	19.85				0	0
50	2	50	2	590	591	10	487-64	19.85 / 19.86	19.85 / 19.86	496-64	19.88 / 19.89	19.85 / 19.86	19.86				0	10
51	3	51	3	580	581	20	485-64	19.20 / 19.22	19.18 / 19.20	539-64	19.19 / 19.20	19.19 / 19.19	19.19				1	19
52	4	52	4	570	571	30	484-64	18.98 / 18.98	18.98 / 18.97	254-63	18.94 / 18.95	18.95 / 18.95	18.96				1	29
53	5	53	5	550	551	50	268-63	19.05 / 19.04	19.09 / 19.08	67465	19.19 / 19.11	19.10 / 19.10	19.09				2	48
54	6	54	6	500	501	100	653-64	18.80 / 18.82	18.83 / 18.84	640-64	18.80 / 18.81	18.84 / 18.84	18.84				3	97
55	7	55	7	450	451	150	637-64	18.72 / 18.74	18.76 / 18.77	642-64	18.79 / 18.80	18.77 / 18.77	18.77				5	145
56	8	56	8	400	400	200	396-64	18.60 / 18.60	18.59 / 18.58	643-64	18.62 / 18.63	18.58 / 18.59	18.58				6	194
57	9	57	9	300	300	300	641-64	17.84 / 17.83	17.85 / 17.84	708-64	17.88 / 17.88	17.87 / 17.86	17.86		291	9	9	291
							WAI No 1	14° DR208°		55-1321	20.63 / 20.64	20.57 / 20.57		2.71				
58	10	58	10	200	200	400	636-64	17.52 / 17.52	17.50 / 17.49	675-65	17.54 / 17.53	17.54 / 17.52	17.51		387	13	12	388
										423-63	21.02 / 21.02	21.01 / 21.00		3.49				
59	11	59	11	100	100	500	510-64	16.92 / 16.91	16.84 / 16.83	157-64	16.81 / 16.82	16.80 / 16.81	16.82		486	14	14	486
							WAI No 2	5° DR210°		419-63	20.96 / 20.96	20.91 / 20.91		4.09				
60	12	60	12	0	0	600	327-64	15.17 / 15.18	15.13 / 15.14	701-64	15.20 / 15.20	15.11 / 15.10	15.12		585	15	15	585
										196-64	20.04 / 20.05	19.96 / 19.97		4.84				

Remarks: 48, 49, 50

FIGURE 12.2
Oceanographic log sheet-A. (Station and thermometer data scanned, cropped, and reprinted from Pub. No. 607, 3rd Edition, 1968.)

Station Preparation

Large oceanographic research vessels cost $2000 per hour (and up) to operate. Being prepared for the upcoming station is clearly a cost-conscious activity. The actual observations to be taken at each station should have been decided in preparing the cruise plan long before the voyage began (Chapter 13). The on-watch scientists and engineers will have reviewed the daily work-plan and insured that the instruments are "prepped" before arriving on station. This means communicating with the ship's crew: the officer on watch on the bridge, the winch operator, the deck crew who might be assisting with swinging the instrument(s) over the side, the computer operator, and the survey technicians.

Before arriving on station, the "science watch" needs to be properly clothed and alert. Short of dangerous seas, the station will be taken, rain or shine, day or night. If the research vessel on which you are working requires certain safety gear, have it on: a stern vest and hard hat are good items on any occasion in any weather. Appropriate footwear goes without saying; decks can be slippery. The weather forecast should always be consulted before arriving on station; conditions can change in the hours a station may take. Be sure a toolbox is handy as certain instruments may need a screwdriver or pliers.

If water samples are to be taken, the sample bottles need to be prepared, numbered, and entered onto the station log. A sample-bottle crate will help in keeping things organized and transported from the deck to the lab. Water samples may be drawn for calibration purposes or for laboratory analysis upon returning to port. If the samples need to be refrigerated or frozen, a samples-only refrigerator/freezer needs to be in the oceanographic lab. Storing samples in the galley "reefers" is never good scientific practice. Cores and bottom grabs may require special handling as may plankton net samples. Chemicals needed for sample preservation have to be ready in the wet lab before arriving on station.

Ancillary Data

If you do not know when and where a sample is taken, you have no useful information. To crystalize the point, consider the surface report of a voluntary observing ship (VOS), which major oceanographic ships are, as well as many commercial vessels. Figure 12.3 details the synoptic report a VOS makes at 0000, 0600, 1200, and 1800 Greenwich Mean Time (GMT) daily. These data are sent by telemetry so as to be available for marine weather reports and forecast model initialization. A common World Meteorological Organization (WMO) format assures that the data are ingested into a database and useful for their intended purpose. The data are also archived and made available for research through the International Combined

Synoptic code symbols with range of values		
BBXX	Ship Weather Report Indicator	BBXX
D....D	Radio call sign	Call Sign
YY	Day of the month	01–31
GG	Time of observation	00–23
i_w	Wind indicator	3, 4
$L_aL_aL_a$	Latitude	000–900
Q_c	Quadrant	1, 3, 5, 7
$L_oL_oL_oL_o$	Longitude	0000–1800
i_R	Precipitation data indicator	4
i_x	Weather data indicator	1, 3
h	Cloud base height	0–9, /
VV	Visibility	90–99
N	Cloud cover	0–9, /
dd	Wind direction	00–36, 99
ff	Wind speed	00–99
fff	High Speed Wind	Knots (099–)
s_n	Sign of temperature	0, 1
TTT	Dry bulb temperature	Celsius Degrees
$T_dT_dT_d$	Dew point temperature	Celsius Degrees
PPPP	Sea level pressure	Actual Hp or Mb (omit 1 in thousandths)
a	3-hour pressure tendency	0–8
ppp	3-hour pressure change	Hp or Mb
ww	Present weather	00–99
W_1	Past weather (primary)	0–9
W_2	Past weather (secondary)	0–9
N_h	Lowest cloud cover	0–9, /
C_L	Low cloud type	0–9, /
C_M	Middle cloud type	0–9, /
C_H	High cloud type	0–9, /
D_s	Ship's course	0–9
V_s	Ship's average speed	0–9
S_s	Sign/type sea surface temp.	0–7
$T_wT_wT_w$	Sea surface temp.	Celsius Degrees
P_wP_w	Sea period	Seconds
H_wH_w	Sea height	Half Meters
$d_{w1}d_{w1}$	Primary swell direction	01–36, 99
$d_{w2}d_{w2}$	Secondary swell direction	01–36, 99, //
$P_{w1}P_{w1}$	Primary swell period	Seconds
$H_{w1}H_{w1}$	Primary swell height	Half Meters
$P_{w2}P_{w2}$	Secondary swell period	Seconds
$H_{w2}H_{w2}$	Secondary swell height	Half Meters
I_s	Ice accretion cause on ship	1–5
E_sE_s	Ice accretion thickness on ship	Centimeters
R_s	Ice accretion rate on ship	0–4
S_w	Sign/type wet bulb temp.	0–7
$T_bT_bT_b$	Wet bulb temp.	Celsius Degress
c_i	Sea ice concentration	0–9, /
S_i	Sea ice development	0–9, /
b_i	Ice of land origin	0–9, /
D_i	Ice edge bearing	0–9, /
z_i	Ice trend	0–9, /

FIGURE 12.3
Marine Surface Weather Observations synoptic code from the *National Weather Service Observing Handbook No. 1* (May 2010 Edition). Contained in the booklet are detailed instructions for entering data into the software and for telemetering the report. *Observing Handbook No. 1* is available from: http://www.vos.noaa.gov/ObsHB-508/ObservingHandbook1_2010_508_compliant.pdf.

Ocean-Atmosphere Data Set (ICOADS), available from the National Center for Atmospheric Research.

US Navy Lieutenant Matthew Fontaine Maury recognized the great value of sharing regular marine weather observations. In 1853, he represented the United States at an international convention in Brussels to standardize such observations (see Figure 12.3). Maury went on to lead the US Navy Depot of Charts and Instruments (now the Oceanographic Office), and was superintendent of the US Naval Observatory. His book *The Physical Geography of the Sea* (1855) is a classic must-read for any seagoing ocean scientist or engineer interested in the history of operational oceanography. Maury's work from studying ship's log books led to publication of *Wind and Current Charts* in 1848, which is a forerunner of today's Pilot Charts (Figure 13.1).

The format of a WMO VOS surface report must read as follows:

BBXX D ... D YYGGI$_w$ 99L$_a$L$_a$L$_a$ Q$_c$L$_o$L$_o$L$_o$L$_o$ i$_R$i$_x$hVV Nddff 00fff 1s$_n$TTT 2s$_n$T$_d$T$_d$T$_d$ 4PPPP 5appp 7wwW$_1$W$_2$ 8N$_h$C$_L$C$_M$C$_H$ 222D$_s$v$_s$ 0S$_s$T$_w$T$_w$T$_w$ 2P$_w$P$_w$H$_w$H$_w$ 3d$_{w1}$d$_{w1}$d$_{w2}$d$_{w2}$ 4P$_{w1}$P$_{w1}$H$_{w1}$H$_{w1}$ 5P$_{w2}$P$_{w2}$H$_{w2}$H$_{w2}$ 6I$_s$E$_s$E$_s$R$_s$ 8s$_w$T$_b$T$_b$T$_b$ ICE c$_i$S$_i$b$_i$D$_i$z (or plain language).

Ship's Synoptic Code Section 0 is mandatory, an example of which might be

BBXX	D ... D	YYGGI$_w$	99L$_a$L$_a$L$_a$	Q$_c$L$_o$L$_o$L$_o$L$_o$
BBXX	WFIT	17123	99280	70805

which is interpreted to mean this is a VOS weather report (BBXX); the ship's call letters are WFIT (not a real ship); the day of the month (YY) is the 17th; the hour of the observation is 1200 GMT; the code for wind speed ($I_w = 3$) means that it is estimated (from the sea state) in knots; the latitude is 28.0°N; the quadrant is $Q_c = 7$ (north latitude, west longitude); and the longitude is 080.5°W. Note that the decimal point for latitude and longitude is omitted in the transmission. (For the curious, this is the latitude [φ] and longitude [λ] of the author's home—see Chapters 7 and 15—and his local National Public Radio station.)

Ship's Synoptic Code Section 1 is in the following table; a numerical example of which is

I$_R$i$_x$hVV	Nddff	00fff	1s$_n$TTT	2s$_n$T$_d$T$_d$T$_d$	4PPPP	5appp	7wwW$_1$W$_2$	8N$_h$C$_L$C$_M$C$_H$
41496	53609	10121	20096	40028	54000	78025	85060

$I_R = 4$ (codes are from *Observing Handbook No. 1*) indicates that no precipitation is reported (standard practice for USA VOS vessels); $i_x = 1$ indicates that group 7wwW$_1$W$_2$ (present and past weather) is included; $h = 4$ shows that the cloud base is 1000–2000 feet; and VV = 96 codes that visibility is 2–5 nm. Group Nddff details cloud cover ($N = 5$ means 5/8th coverage); wind direction (dd) ranges from 00 for calm to 36 (meaning a north wind); wind speed (ff) is from the Beaufort Scale (Figure 12.4)

Beaufort Number	Descriptive Term	Knots	Specification	Probable Wave Height (feet)	Wind Symbol
0	Calm	0	Sea like a mirror; oil slicks; smooth swell	0	
1	Light Air	1–3	Ripples with the appearance of scales; catspaws	1/4	
2	Light Breeze	4–6	Small wavelets; crests have a glassy appearance	1/2	
3	Gentle Breeze	7–10	Large wavelets; crests begin to break; scattered white caps	2	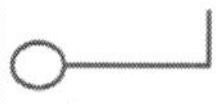
4	Moderate Breeze	11–16	Small waves; frequent white caps; few spray	3 1/2	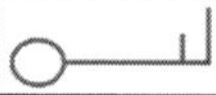
5	Fresh Breeze	17–21	Moderate waves; many white caps; few spray	6	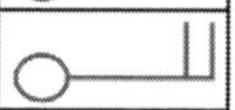
6	Strong Breeze	22–27	Large waves begin to form; extensive foam crests; some spray	91/2	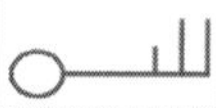
7	Near Gale	28–33	Foam from breaking waves begins to be blown in streaks	13 1/2	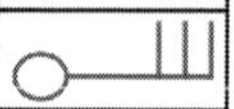
8	Gale	34–40	Moderately high waves; foam blown in well-marked streaks	18	
9	Strong Gale	41–47	High waves; dense foam streaks; spray may affect visibility	23	
10	Storm	48–55	Very high waves; great patches of foam; visbility affected	29	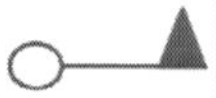
11	Violent Storm	56–63	Exceptionally high waves; sea completely foam covered; long patches of foam	37	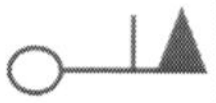
12	Hurricane	64+	Air filled with foam; sea completely white; visibilty seriously affected	45	

FIGURE 12.4

Beaufort scale of wind, waves, and sea state based on verbiage in *Observing Handbook No. 1.* Wind symbols are standard meteorological usage—short barb is 5 kn, long barb is 10 kn, and the flag is 50 kn. If the wind speed were (say) 125 kn, the symbol would be two flags, two long barbs, and one short barb.

and ranges from 00 for calm to 12 for hurricane speeds (64 kn and over). ff = 09 is a gentle breeze of 7–10 kn. In this example, 00fff would be omitted as the wind is less than 99 kn. In group $1s_nTTT$, the zero signifies the (dry-bulb) air temperature is positive or zero, and TTT = 121 is +12.1°C. Dew point temperature $(T_dT_dT_d)$ is similarly coded for +09.6°C. The number 40028 indicates that sea level pressure is 1002.8 mb (1 mb = 1 hPa). Group 5appp is coded such that $a = 4$ indicates pressure tendency is steady with no change (ppp = 00.0 mb). From tables in *Observing Handbook No. 1*, ww = 80 reports a slight rain shower, W_1W_2 = 25 is cloud cover more than half throughout the reporting period, and W_2 = 5 indicates drizzle. Finally, $8N_hC_LC_MC_H$ reports low, medium, and high clouds present; $N_h = 5$

reiterates that 5 eights are medium (M) level clouds, with $C_M = 6$ meaning altocumulus from the spreading of cumulus clouds.

Units in these observations are a hybrid system, mixing classical measures of nautical miles, knots, hours, yards, and feet with Celsius degrees, hectopascals or millibars, meters or kilometers, and seconds. By using the codes, the units are interchangeable. For example, VV = 96 (earlier paragraph)—then the visibility is 4–10 km or 2–5 nm; if VV = 91 were reported, it means that the visibility is between 50 and 200 m or 55 and 220 yards (1 m = 1.1 yards), and so forth. Wind direction is always reported as the direction *from* which it is blowing, and is reported in degrees true—that is with respect to due north. If apparent wind is observed it must be converted to true wind before being reported.

Ship's Synoptic Code Section 2 is the surface oceanography data section of *Observing Handbook No. 1*. The nine codes, $222D_s v_s$, $0s_s T_w T_w T_w$, $2P_w P_w H_w H_w$, $3d_{w1} d_{w1} d_{w2} d_{w2}$, $4P_{w1} P_{w1} H_{w1} H_{w1}$, $5P_{w2} P_{w2} H_{w2} H_{w2}$, $6I_s E_s E_s R_s$, $8s_w T_b T_b T_b$, and ICE $c_i S_i b_i D_i z_i$ are discussed as follows. Consider the first six codes in Section 2 (*Observing Handbook No. 1*):

$222D_s v_s$	$0s_s T_w T_w T_w$	$2P_w P_w H_w H_w$	$3d_{w1} d_{w1} d_{w2} d_{w2}$	$4P_{w1} P_{w1} H_{w1} H_{w1}$	$5P_{w2} P_{w2} H_{w2} H_{w2}$
22233	02207	20802	309//	41204	5////

$D_s = 3$ is a code to mean that the ship's true course is southeast, and $v_s = 3$ means that she is going 11–15 kn. $S_s = 2$ shows that the sea surface temperature ($T_w T_w T_w = 207$) is a positive bucket thermometer measurement of 20.7°C. The locally generated wind wave period $P_w P_w$ is 8 seconds and the significant wave height is coded as $H_w H_w = 02$ meaning the waves are 3 or 4 feet (~1 m) in height. The next code quantifies the direction of primary swell waves (as distinct from $2P_w P_w H_w H_w$, which is the local wind waves); in this example the primary swell is from 090° and // indicates that there is no secondary swell. Group $4P_{w1} P_{w1} H_{w1} H_{w1}$ is used to describe the primary swell period and significant wave height; in this example 12 is the period in seconds, and $H_{w1} H_{w1} = 04$ from the code book is for a 6 or 7 feet swell. Code 5//// confirms that there are no secondary swell, and group "5" would not be reported because of group $3d_{w1} d_{w1} d_{w2} d_{w2}$ using slashes.

Group 6 details the cause of ice accretion on the ship and is not reported in this example as the research vessel is in the offing of Cape Canaveral, Florida; $E_s E_s$ and R_s are coded for thickness of ice accretion and rate of ice accretion, respectively (if needed). Group 8 is for reporting the sign and method of determining wet bulb temperature; an example, $8s_w T_b T_b T_b = 80101$, reports that wet-bulb temperature is 10.1°C and is positive ($s_w = 0$). Wet-bulb temperature and air temperature (dry-bulb temperature) are measured with a sling psychrometer. Group $c_i S_i b_i D_i z_i$ is for reporting sea ice and ice of land origin; the codes allow quantifying the concentration of sea ice (c_i), the sea ice stage of development (S_i), icebergs with growlers and bergy bits (b_i), and the bearing of the principal ice edge (D_i). Finally, z_i codifies information such as the sea ice situation (ship in distress?) and the 3-hour trend.

As can be seen, *Observing Handbook No. 1* is a very thorough reporting method for surface observations. All VOS, and especially research vessel observations, form a critical link in weather forecasting and climate documentation. On a cautionary note, the sea surface temperature ($T_wT_wT_w$) from a bucket thermometer is in general much preferred to that taken from a sea chest (Figure 3.3); the depth of the sea chest depends on the ship's draft, and can be 50 feet below the sea surface on a large vessel. In all cases, the instruments must be calibrated, and the National Weather Service (NWS) has established a team of Port Meteorological Officers (PMO) who visit ships and post corrections to instruments such as the aneroid barometer.

Voluntary Observing Ship Digital Input

Much of the information discussed before is now input directly into a computer file. Figure 12.5 shows the screen that opens when the VOS computer program is started. The program, "TURBOWIN Plus" is distributed by NOAA through the local PMO. PMOs are highly educated professional marine meteorologists that visit ships while in port, train the observers, calibrate instruments—the barometer in particular—and offer advice on how to make better observations and the value of participating in the program (all ships benefit from these observations, which are the initial conditions essential for marine weather forecasts).

Each of the terms on the upper toolbar, "File," "Input," "Output," and so on, have drop-down menus with the typical information needed to make decisions. The system has multiple paths to entering data. For example, clicking on the word "Input" drops a menu with: Date and Time..., Position, Course and Speed..., Barometer Reading, Barograph Readings..., Temperatures..., and so forth. Words on the form itself (Figure 12.5, left-hand column) are all active too; passing the computer mouse over any one of them opens another window for entering data. Each of the icons on the lower toolbar also allow opening secondary windows for entering data. The toolbar is reproduced in Figure 12.6.

Contained in drop-down input menus are photographs of clouds, barograph tendency, and other details contained in *Observing Handbook No. 1*. The computer program is not so complete that *Observing Handbook No. 1* is completely superseded. Information such as the Beaufort scale (Figure 12.4) for estimating sea state is still the preferred method of estimating wind speed, too, because anemometers often are influenced by ship-generated air turbulence. The computer program is set up so that clicking "OK" advances the program to the next field to be entered in the same sequence as shown in Figure 12.6.

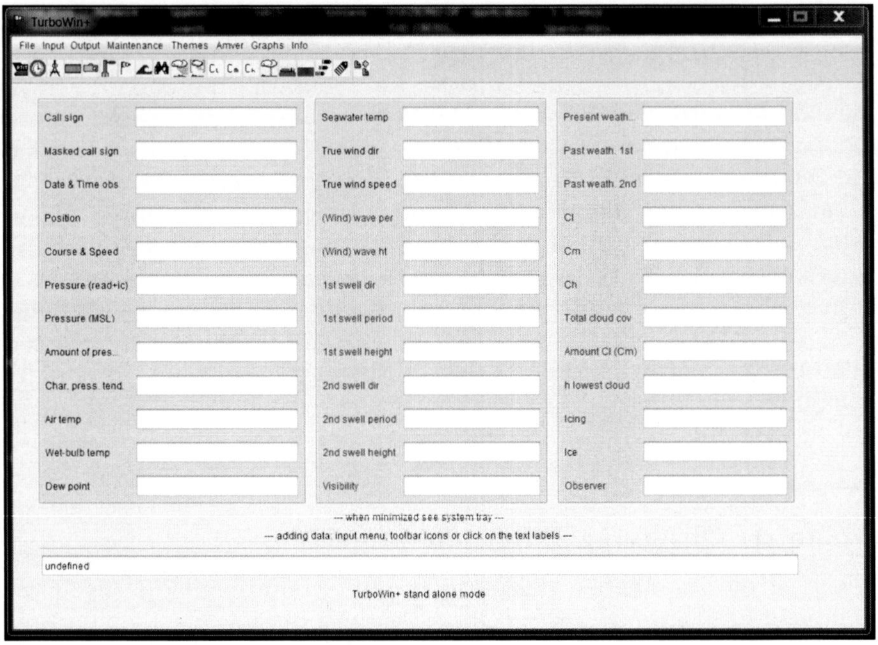

FIGURE 12.5

Voluntary observing ship (VOS) reporting form automating World Meteorological Organization (WMO) international format. (See http://www.vos.noaa.gov/ for details.)

FIGURE 12.6

Icons from lower toolbar in Figure 12.5. From left to right: Bow = call letters of vessel, Clock = time, Dividers = position, Box = barometer, Next box = barograph, Thermometer = air temperatures, Flag = wind, Wave = sea state, Binoculars = visibility, Cloud = present weather, Bracketed cloud = past weather, C_L = low clouds, C_M = middle clouds, C_h = high clouds, Cloud arrow = cloud height, Ship on blue = ice accretion, Blue box = ice, Shoulder boards = observers, Four stripes = captain's name, and Plus sign = course and speed (repeat of dividers).

Apparent Wind → True Wind

The most egregious mistake in the ICOADS and WMO datasets is reporting wind direction. Harken back to the last century. Mike Uporski is the chief quartermaster (CQM—better look that up!). A very "green" Ensign is the Officer of the Deck (OOD) (better look that up too—see Chapter 2) and is recording the 0300 weather on the "mid-watch." CQM Mike asks if the wind is true or apparent. This "green" Ensign knows the difference, and shows the CQM his plot on a maneuvering board.

Although $D_s = 3$, the true course is 145° (see Figure 7.4) and the actual speed is 12 kn. The OOD observes from the anemometer that the apparent wind relative to the ship's bow is 312° at 7 kn, and thus is 312° + 145° = 097° at 7 kn relative to the ship's course. The OOD plots the solution on the maneuvering board (Figure 12.7) showing that the true wind is from 000° (due north) at 9 kn. Thus, the entries for dd = 36 and ff = 09 in Nddff are reported. The seagoing scientist or engineer would check these results from the trigonometric law of cosines and sines solving $c^2 = a^2 + b^2 - 2ab \cos C$ and $\dfrac{\sin A}{a} = \dfrac{\sin C}{c}$, respectively, where capital "A" and capital "C" are the angles opposite sides "a" and "c." In this example, $C = 145° - 097° = 48°$, $a = 7$, and $b = 12$.

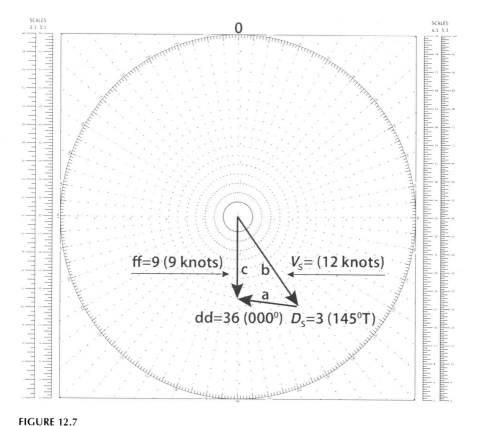

FIGURE 12.7
Maneuvering board graphical solution for vessel on course 145° T (code $D_s = 3$) at 12 kn (code $v_s = 3$) with an apparent wind from 097° at 7 kn relative to the ship's course. The true wind is from the north (code dd = 36) at 9 kn (code ff = 9). The concentric circles are labeled 1–10, and are doubled for plotting this example.

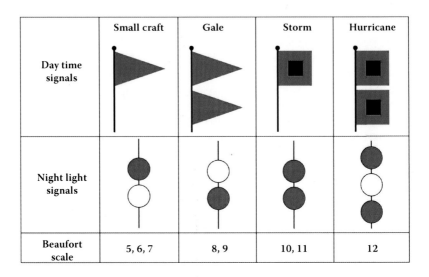

FIGURE 12.8
Daytime flag signals and nighttime light signals to inform mariners of offshore conditions. These signals are usually displayed by the harbormaster and are in plain sight to vessels leaving the port.

The conscientious observer always will check the maneuvering board solution by looking at the sea and insuring that the sea state and direction from the ship's compass agrees with the maneuvering board solution.

In the marine surface weather observation synoptic code example, the wind speed (ff) was determined from the Beaufort scale (Figure 12.4) to be a three (3), and is confirmed with the maneuvering board solution shown in Figure 12.7. This surface weather scale was developed in England by Sir Francis Beaufort in 1805, and is used on land and at sea. Before leaving port, weather conditions and forecasts must be obtained. Classically, a flagpole in the harbor will have lights and/or flags (Figure 12.8) displayed to inform mariners of the current weather. Small craft warnings are particularly important for 40- to 65 foot LOA research vessels as waves from 6 to 13 feet may be encountered. Even a large vessel would not be advised to enter the sea with gale force winds (Beaufort 8–9) offshore.

"The good seaman weathers the storm he cannot avoid, and avoids the storm he cannot weather."—Author unknown, weather proverb.

Additional Reading

Hayler, W.B., J.M. Keever, and P.M. Seller. 2003. *American Merchant Seaman's Manual*, 7th Edition. Cornell, NY: Maritime Press, 672 pp.

Exercises

1. Plan a series of ten oceanographic stations equally spaced between Fowey Rocks and Gun Cay, Bahamas, across the Straits of Florida. Each conductivity, temperature, depth (CTD) cast is to go to within 10 m of the seafloor. Find the largest scale nautical chart of the crossing, space all stations, tabulate the latitudes and longitudes, estimate water depths, estimate cast time at 50 m per minute on the CTD winch, and list ancillary data as a VOS. Make an instrument list.

2. Download *Observing Handbook No. 1,* and create a PowerPoint presentation to train your scientific party in how to make the VOS observations. Create a series of observations expected for the time of year your cruise in Question 1 is to take place. List the instruments needed and demonstrate taking the surface observations.

3. "What is it?"—think calculations...

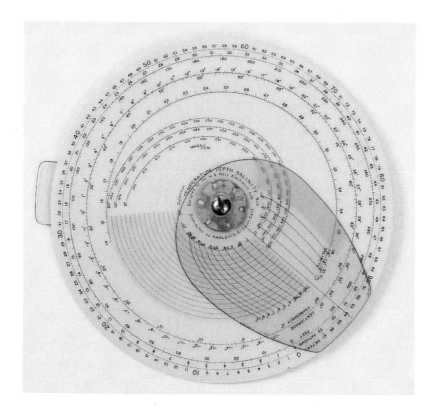

13

Cruise Planning and Execution—Project Instructions, Foreign Waters, Geographic Names, and Tides and Tidal Currents

Proper Planning Prevents Poor Performance—the five P's of readying for an oceanographic expedition (read that before?). As with any venture to sea, the planning stage is the essential first step. Written *project instructions* are the method of proper planning. Writing focuses the process and quantifies the resources needed for a successful voyage. However, even before starting to write, the seagoing scientist or engineer needs to research the environmental conditions that likely will be encountered in the area of operations. Thanks to Matthew Fontaine Maury and other nineteenth century visionaries, such information is available in Pilot Charts (Figure 13.1).

The information on the Pilot Chart includes month-by-month air and water temperatures, sea level (barometric) pressure, main tracks of tropical cyclones and hurricanes, percent of ship reports with Beaufort Force 8 or higher winds (see Chapter 12), iceberg and growler locations of note, sea ice limits, magnetic variation (see Chapter 6), percentage of observations showing a visibility of 2 nm or less, probability of wave height exceeding 12 feet, ocean current speed and direction, and wind roses that document the percentage of winds from cardinal (N, E, S, and W) and intercardinal (NE, SE, SW, and NW) directions. Time spent studying the Pilot Chart will guide the oceanographer or ocean engineer as to the expected conditions to be encountered during their planned voyage, and watching daily weather reports and forecasts will add to the appreciation of the variability to be encountered.

Pilot Charts are one of the many specialized Maritime Safety Information (MSI) publications freely available at the National Geospatial-Intelligence Agency (NGA) website (http://msi.nga.mil/NGAPortal/MSI.portal). Their list of publications includes: *American Practical Navigator, Atlas of Pilot Charts, Chart No. 1* (see Chapter 7), *Distance between Ports, International Code of Signals* (see Chapter 3), *NGA List of Lights* (foreign waters), *Radio Navigation Aids, Sailing Directions Enroute, Sailing Directions Planning Guide, Sight Reduction Tables for Marine Navigation* (see Chapter 15), *USCG Light List* (US waters), and *World Port Index*. MSI data are kept up-to-date with *Notices to Mariners*. In addition, the National Oceanic and Atmospheric Administration publishes nautical charts, tide tables, tidal current tables, distances between US ports, and the very valuable *Coast Pilot*. Most of these products and data are also

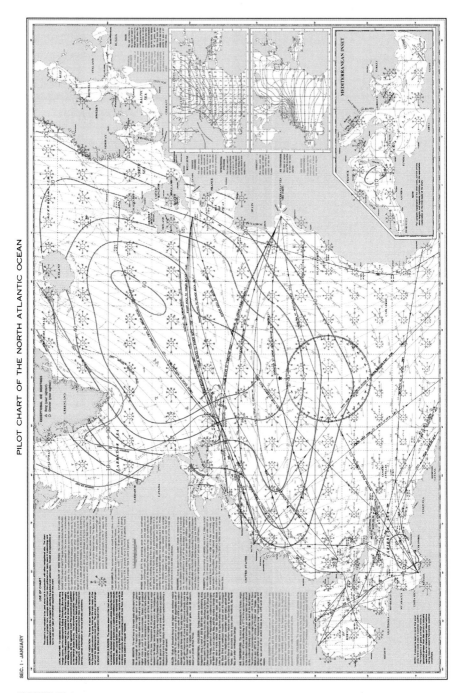

FIGURE 13.1
Pilot chart of the North Atlantic Ocean for January. The red contours show the probability of waves exceeding 12 feet. Wind roses show prevailing winds with percent calm in the circle. Black lines show major shipping routes. Green arrows show surface current speed and direction.

freely available online at http://tidesandcurrents.noaa.gov/products.html, including historical *Tide Tables*, and *Tidal Current Tables*. In addition, the US Naval Observatory publishes *The Nautical Almanac* for celestial navigation, and other related products.

Many warning products from the NGA are time-sensitive for the protection of vessels and safety of life at sea (SOLAS) (see Chapter 4). Examples are US Office of Naval Intelligence reports on acts of piracy, anti-shipping activity messages, HYDROARC (NGA hydrographic product for the Arctic) reports of hazards in international waters of the Arctic Ocean, and worldwide navigational warning service (WWNWS) broadcasts of immediate hazards to navigation. Research vessels of course contribute to such information through MSI reports. In coastal waters of the United States, such information may be broadcast on channel 16 (VHF 156.8 MHz), where the information is received by many ships and the US Coast Guard. Many research vessels will also have a single-side-band radio set to 2182 kHz which offers much longer range than channel 16, but serves a similar purpose. Channel 16 and 2182 kHz are international calling and distress frequencies for maritime communication.

The seagoing scientist or engineer will not need to consult all of the publications or services mentioned as earlier, but he/she should be cognizant of them. Safe navigation is the responsibility of the research vessel's captain and deck officers, but in planning and writing project instructions, the chief scientist and the scientific party have to make judgments as to whether the observations planned are practicable in the expected weather conditions. The final authority while at sea as to the safety of an operation rests with the captain.

Project Instructions

Project instructions should be written in a format similar to any technical publication. Not only is the writing a five P's (proper planning prevents poor performance) exercise, it informs the captain and crew of the purpose for the voyage and alerts them to special requirements. It also informs the scientific party of the work to be done. Such a document should have the following format:

- **Title**: Authors(s) and their professional affiliation.
- **Abstract**: Provide here a one-paragraph succinct summary of the work to be accomplished and why. Put the planned dates here and below so that timely permissions can be obtained if an international voyage of discovery.
- **Introduction**: In this section, what the project is about is written for the captain, officers, and crew of the research vessel, as well as the

scientific party. The crew may not be scientists, so keep it understandable; they are not uneducated or inexperienced, so do not patronize.

- **Personnel**: A list of the scientific party and their assignments is given. Are there any foreign nationals in the scientific party? Are there any special-needs scientists or technicians? Dietary restrictions? Are passports required (*yes* if an international voyage)?

- **Observations**: In this section, describe the data that are being acquired and the methods of acquiring it. Add a brief mention of the data acquired in the past, with perhaps a brief historical reference. Use a proper (formal) writing format.

- **Equipment**: Here the scientific equipment is listed in detail. Be sure to separate the equipment that you will be bringing from the equipment that you expect to be on board. The captain and chief engineer need to know what deck machinery is needed, if crewmen will be handling heavy items or difficult to handle objects, and so on. Will samples be refrigerated or frozen, and brought back or shipped back? Will any radioactive materials be brought on board? Are the electrical connectors on the hydrographic winch compatible with the instruments being used? Other considerations?

- **Cruise Plan**: A list of the stations including the latitude, longitude, and water depth is required; approximate dates should be listed too. What data are being taken at each station must be listed. Is the order of taking stations critical? If water samples are to be collected from a conductivity, temperature, depth rosette, the rosette depths must be listed. Suggest ports of call, suggest routes from point to point, but remember that the safety of the ship and personnel are the captain's responsibility; respect that burden and authority. Suggest a regular daily (or more often) meeting time; keep communications open and flowing: before, during, and after the voyage.

- **Conclusions**: The project instructions could be written according to the format of *Geophysical Research Letters* (GRL). Note that 12-point New Times Roman font is recommended. Try to italicize Latin terms like *etc.*, *q.v.*, *et al.*, *e.g.*, and/or *i.e.* Be sure to keep their meaning straight.

- **References**: If you cited scientific or engineering literature in the project instructions (encouraged), list sources alphabetically in standard format (GRL is recommended as it is universally recognized, and will allow easy access both for the crew and the scientific party).

Exclusive Economic Zone

From about 30 BC on, the Roman Empire called the Mediterranean Sea *mare nostrum*, "our sea." The notion of restricting access more likely dates to the middle ages when Genoa and Venice were republics; *mare clausum* came to be. Nations were protecting fishing rights and freedom of navigation; legal battles and battles with canon ensued. In 1702, Dutch jurist Cornelius van Bynkershoek proposed that a three-mile zone surround maritime nations over which they had control. The three-mile limit, perhaps related to the range of cannon fire or perhaps related to the league (3 nm at sea in England), was the "territorial sea" until 1945 when Harry S. Truman declared the continental shelf to be in US waters. Other countries followed suit, and this wrangling led to 1982 with the United Nations Convention on the Law of the Sea (UNCLOS), which proposed a 12 nm Territorial Sea, a 24 nm Contiguous Zone, and a 200 nm Exclusive Economic Zone (EEZ). Now about 40% of Earth's oceans are contained within the EEZ of nations; the EEZ of the United States is shown in Figure 13.2.

The waters of a country's EEZ are restricted even to scientific research. The chief scientist planning a cruise is responsible for obtaining permission from

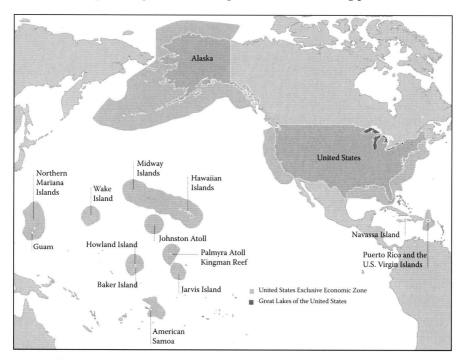

FIGURE 13.2
Exclusive Economic Zone of the United States. The horizontal distance is measured from the "baseline," which is the intersection of mean low water (MLW) with the land. The "shoreline" on the other hand, is the intersection of mean high water (MHW) with the land (see Figure 7.3).

the country whose waters are in the project instructions. Such permission for a United States flagged ship *must be* initiated through the US Department of State using the Research Application Tracking System (RATS) (http://go.usa.gov/3mAMm). A foreign country must approve or deny such permission within 6 months of the application; generally most countries will do so within 3 months.

Under UNCLOS, the "coastal state" (country with sovereignty over the EEZ) has the right to have an observer on board. This may mean a port call to embark the observer, and another port call to debark the observer. It is strongly recommended that a colleague in the coastal state be invited during the planning stage, and be named explicitly in the RATS documents. Research vessels taking routine marine weather data (see Chapter 12), and who are reporting through the Voluntary Observing Ship program, do not have to have permission from the coastal State whose EEZ waters are being transited. The United States is not a signatory nation to UNCLOS.

Geographic Names

"What is in a name?" (William Shakespeare, *Romeo and Juliet*, Act 2). If it is a geographic name, then in the United States it must be approved by the US Board on Geographic Names, an agency within the Department of the Interior. NOAA nautical charts carefully use only the approved geographic names, and seagoing scientists and engineers are encouraged to do so too, as educated men and women should.

Over time geographic names change, such as when a country splits and becomes two separate States. The geographic name in one language may be different from that in another language. An interesting tidbit of name changing was the renaming of Cape Canaveral to Cape Kennedy through a presidential executive order by Lyndon B. Johnson in 1963. The US Board on Geographic Names demurred the same year. Floridians have called the area Cape Canaveral for over 400 years, and were not pleased with Mr. Johnson's action; accordingly in 1973 the Florida legislature put into law that all official state maps and documents would use the historical name. Later that same year, the Interior Department agency changed the name again, but for 10 years it was officially Cape Kennedy.

Tides

Gravitational tides of the ocean were first successfully forecast by William Thompson (Lord Kelvin) *ca.* 1867. Today, almost any laptop computer is capable of making tidal forecasts, the knowledge of which is central to cruise planning in coastal waters. Deep draft vessels in particular must pay attention to the stage of the tide for safe navigation—water depths as well as overhead clearances (see Chapter 7). Water levels at a tide gauge change with the seasons due to nongravitational effects too, such as winds, thermal expansion and contraction, atmospheric barometric pressure, and coastal currents. Interannual effects such as El Niño and long-term trends such as climate change and vertical land motion also are recorded, but not predicted in general.

Gravitational tides are the result of the Earth–Sun–Moon system's motion. Figure 13.3 shows that it is the barycenter of the Earth–Moon system that forms the elliptical orbit about the Sun. The barycenter (think of it as the fulcrum of a lever) is the center of mass and is readily calculated from $M_E r = M_M (R - r)$, where Earth's mass $M_E = 5.97 \times 10^{24}$ kg, Earth's radius $a = 6378$ km, the Moon's mass $M_M = 7.35 \times 10^{22}$ kg, and the mean Earth–Moon distance $R = 384,400$ km; substituting gives $r = 4675$ km or $a - r = 1703$ km below Earth's surface. Thus, Earth's center forms a sinusoidal path about the smooth solar orbit.

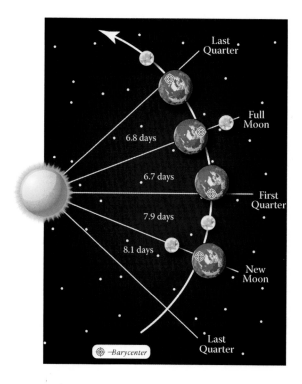

FIGURE 13.3
Earth–Sun–Moon system. Earth and its Moon revolve around a common center of mass called the "barycenter." The motion is similar to that of one's hand twirling a rock on a string in the horizontal plane—all points on the hand have the same radius of motion. The barycenter forms the smooth ellipse known as Earth's solar orbit. Earth's center is a sinusoid around the ellipse.

The frequency (ω) about the barycenter is just the balance between gravitational and centrifugal forces: $G\dfrac{M_E M_M}{R^2} = M_M \omega^2 (R - r)$, where $G = 6.674 \times 10^{-11}\ \mathrm{N} \times \mathrm{m}^2\ \mathrm{kg}^{-2}$ is the familiar universal gravitational constant. A bit of arithmetic, recalling that period $T = \dfrac{2\pi}{\omega}$, gives the sidereal month $T = 27.32$ days, one of the six fundamental periods of gravitational tides.

The tide-producing force is the difference between the centrifugal force and the gravitational force on Earth's *surface*. With the motion being about the barycenter, the gravitational force by the Moon at Earth's center is exactly balanced by the centrifugal force at Earth's center, and because all points on Earth have the same $\omega^2 r$, the centrifugal force is the same everywhere, that is, $M_E \omega^2 r = \dfrac{GM_E M_M}{R^2}$. That is: $F_{\text{tide}} = \left[F_g\right]_{\text{surface}} - \left[F_c\right]_{\text{everywhere}}$ along a line between Earth's center and the Moon's center. Substituting, $F_{\text{tide}} = \dfrac{GM_E M_M}{(R-a)^2} - \dfrac{GM_E M_M}{R^2}$, and with a bit of algebra: $F_{\text{tide}} \approx \dfrac{2aGM_E M_M}{R^3}$. Notice that the tide-producing force is the motion about the barycenter during the sidereal month. Yes, there are generally two high tides and two low tides a day, but Earth's rotation only *times* the tides—it does not *force* the tides.

There are two tidal bulges on Earth, one on the side toward the Moon, where gravitational attraction exceeds centrifugal force, and on the opposite side of Earth where the centrifugal force exceeds the gravitational force. As Earth rotates on its axis, these tidal bulges are passed twice each lunar day ($24^h 50^m$). Hence there are generally two high waters and two low waters each lunar day. Thus the principal semidiurnal lunar tidal constituent, abbreviated M_2, is experienced in most coastal sites open to the deep sea. So, it is the revolution of the Earth–Moon system about the barycenter every 27.32 days that *forces* the tides, and the rotation of Earth about its polar axis that *times* the tides every 12.42 hours or so.

The sidereal month is the motion of the Earth–Moon system with respect to the stars. Since it takes a bit of extra time to align with the sun as the barycenter moves along the solar orbit, the synodic month is 29.53 days. Thus, new moon to new moon is 7.9 days + 6.7 days + 6.8 days + 8.1 days (Figure 13.3), or the familiar 29 ½ days. At syzygy, when the Earth–Moon–Sun are aligned, higher high tides and lower low tides are experienced. Since this happens twice a month, spring tides with higher highs and lower lows occur every 14.76 days, and neap tides with higher lows and lower highs also occur every 14.76 days when the Moon is in quadrature. Figure 13.4 is an example of these monthly cycles.

The abbreviations in Figure 13.4 are LHW, lower high water; HHW, higher high water; LLW, lower low water; HLW, higher low water; and MTL, mean tide level. The chart datum for this site would be mean lower low water (MLLW), which as can be seen, would provide the safest water

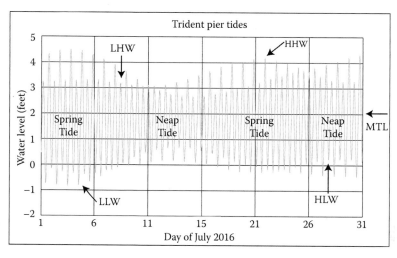

FIGURE 13.4

A month of hourly heights in Port Canaveral, Florida. The tide is predicted from the gravitational harmonic constants and is dominated by the M_2 (principal lunar semidiurnal constituent; 12.42^h), S_2 (principal solar semidiurnal constituent; 12.00^h), N_2 (larger lunar elliptic constituent; 12.66^h), K_1 (luni-solar diurnal constituent; 23.93^h), and O_1 (principal lunar diurnal constituent; 25.82^h). A tidal form factor is often calculated from $\dfrac{M_2 + S_2}{K_1 + O_1}$ amplitudes.

depth because a higher stage of the tide would offer more depth. Note also in Figure 13.4 that the neap tide near day 11 is rather different from that near day 26. Although most of the spring-neap cycle is the interaction of the principle lunar semidiurnal constituent (M_2) and the principle solar semidiurnal constituent (S_2), that is, $\dfrac{1}{12.00^h} - \dfrac{1}{12.42^h} = \dfrac{1}{14.76^d}$ (do the arithmetic...), other tidal constituents in the tidal height (h) prediction equation

$$h(t) = \text{MTL} + \sum_{i=1}^{i=n} A_i \cos(\omega_i t - \varphi_i)$$ add up at a given time (t); as many as $n = 37$

constituents are used in the United States, where A_i is the amplitude, ω_i is the frequency, and φ_i is the phase of the ith constituent, say that of the M_2 with a period of $12^h 25^m$.

Tidal height predictions $(h[t])$ are available from the NOAA Center for Operational Oceanographic Products and Services (http://tidesandcurrents.noaa.gov) online. Older data are in *Tide Tables*, often found in major libraries in coastal cities and universities. For foreign sites, such as England, the United Kingdom Hydrographic Office is an excellent resource. *Tidal Current Tables* and predictions are also available from these national sources.

Tidal Currents

Tides are shallow water waves whose celerity (c) is given by $c = \sqrt{gZ}$, where Z is the water depth, and g is gravity. The ocean is not deep enough for tides to be free waves, and continents and bottom topography force the tidal waves to conform to boundaries. In many areas, tidal waves behave as progressive waves where the celerity is a maximum at the crest and trough, and in other areas they are standing waves where the maximum celerity is at the halfway point between the crest and trough. In an ocean area such as the North Atlantic, the tidal wave is mathematically described as a Kelvin Wave. In the North Atlantic's M_2 Kelvin Wave, the range of the tide is greatest at the coasts, and the wave sweeps anticlockwise around the amphidromic point of the basin as a shallow water wave.

For many research vessel operations in coastal areas, and at inlets in particular, the tidal currents behave much like that of a hydraulic difference between the height of the tide in the open ocean and the height inside the inlet, in a bay or lagoon. Figure 13.5 is an illustration of such a tidal current $U(t)$ through an imaginary inlet. Slack water, when the current is zero, occurs when the water level in the ocean is the same as in the bay. Three such events are shown in Figure 13.5; maximum ebb current occurs at about $t = 150°$, and maximum flood current at about $t = 330°$. Nothing in this illustration is to suggest the magnitude of the tidal current, just the phase relationship to the height difference between the ocean tide and the bay tide. The simplest model for the magnitude would be a balance of the pressure gradient force and friction: $g\dfrac{\partial h}{\partial x} = JU$, where J is the linear Guldberg–Mohn friction coefficient, and $\dfrac{\partial h}{\partial x}$ is the slope of the sea surface due to the tidal height difference.

In practice, tidal currents are measured at locations in a harbor or estuary by current meters. The Roberts Radio Current Meter (see Figure 4.5) was deployed in many locations critical to safe navigation in the United States. Observations were taken near the surface, at mid-depth, and near the bottom, whereas modern acoustic Doppler current meters can profile the entire water column. From the current-profile data, the amplitude and phase of the maximum flood current and ebb current are determined by harmonic analysis, and predictions are created. *Tidal Current Tables*, which also contain the times of slack water, are mostly replaced by websites such as the NOAA site (https://tidesandcurrents.noaa.gov/curr_pred.html). Seagoing scientists and engineers need to know current speed and direction in planning and executing research vessel operations in or near tidal streams. Tidal currents of 2–3 kn are not uncommon, reverse direction every 6.21 hours, and must be accounted for in safe operations. Saltstraumen on Norway's northwest coast boasts the world's record tidal current: 22 kn!

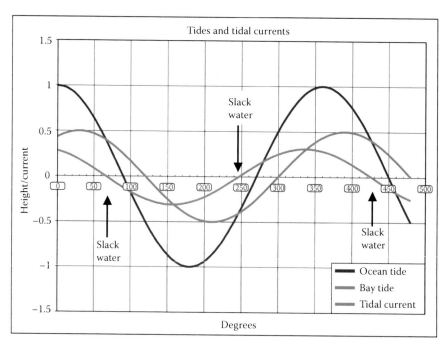

FIGURE 13.5

Simplified visualization of the relationship between the tidal current and the height difference of the tide in the ocean and in the juxtaposed bay. Amplitude of the ocean tide is $A_{ocean} = 1$; amplitude of the bay tide is $A_{bay} = 0.5$; phase of the ocean tide is zero, but the phase of the bay tide is $\varphi_{bay} = 30°$. Using $h(t) = A\cos(\frac{2\pi}{360}t - \varphi)$ over the interval $0° > t > 500°$, the two tidal curves are plotted and the tidal current U is calculated from $U(t) = h_{ocean}(t) - h_{bay}(t)$.

Away from inlets, the tidal currents are called rotary tidal currents, and in general they are much weaker. They are driven by the passing of the Kelvin Wave, and in a semidiurnal tide such as the M_2, rotate through 360° in 12.42 hours. The speed and direction of rotary tidal currents ranges from 0.1 to 0.5 kn on the east coast US continental shelf; they have a distinct lack of slack water, and rotate clockwise. Rotary tidal current information is found in Table 5 of *Tidal Current Tables*. For operational planning, especially along the continental shelf break, the strength and direction of offshore currents must be factored in. Between Key West and Cape Hatteras for example, the Gulf Stream can easily top 3 kn, and is complicated by countercurrents, streamers, filaments, and eddies. Extreme caution must be exercised during underwater operations as boats and divers can be swept away rapidly.

Additional Reading

Van Dorn, W.G. 1974. *Oceanography and Seamanship*. New York, NY: Dodd, Mead & Co, 481 pp.

Exercises

1. For the cruise in Question 1 of Chapter 12, write "project instructions" using the format: Title, Author, Abstract, Introduction, Personnel, Observations, Equipment, Cruise Plan, Conclusions, and References. You will be entering foreign waters (the Commonwealth of the Bahamas); what additional effort must you initiate and with whom? Plan your cruise for the research vessel in Figure 3.1.
2. Find tide information for departing Fowey Rocks and arriving in Gun Cay, The Bahamas. When is high tide? When is low tide? If you need information beyond the usual three-day forecast, where is it available?
3. The box is about 28 inches long and 14 inches high. What is it?

14

Underwater Operations—Planning, Equipment, Safety, Diving, American Academy of Underwater Sciences Certification, and Underwater Archeology

Mid-watch on the USC&GS Ship *Discoverer*, 0200 actually, travelling from Boston to Jacksonville. Rang up "half ahead" on the engine room telegraph; time to deploy the magnetometer. Chief Survey Technician Chuck Ellis wraps the magnetometer cable three turns around the 24-inch diameter capstan and feeds the "fish" out the chock on the fantail. Water depth 25 fathoms; do not want the instrument to hit bottom, but 600 ft is the preferred amount of cable out to avoid the magnetic signature of *Discoverer* on the geomagnetic observations. Chaffing gear at the ready; magnetometer deployed. All ahead "full"; engine room replies "full ahead." Time to change course approaching buoy "1"; right 10° rudder; come to 090. Looking aft at the wake in the moonlight, the buoy was passed close to starboard; will the magnetometer snag on the mooring? Anxious minutes go by at 12 kn. Made it safely, but next time the Officer of the Deck (OOD) will know better and give the obstruction a wider berth!

Proper planning prevents poor performance—saw that before, correct? Towing an instrument is an underwater operation, be it a magnetometer underway, or a bottom camera while drifting over a seamount 300 fathoms below, or a CTD cast to near the seafloor in the Gulf Stream. The operation requires teamwork: winch operator, boatswain, technician, scientist, engineer, deck watch, engine room watch, captain.... Before putting devices overboard, the environment must be understood: What is the weather forecast? What is the water depth? Are there any strong currents? Is the crew briefed and practiced? Is the equipment prepped and ready before arriving on station? Are all safety precautions taken? Are swimmers or divers required? Are all systems go?

Each underwater operation should have a checklist. A ship in the open ocean is subject to motions (see Figure 4.1) that challenge the footing of even the most experienced seagoing scientist or engineer. Working under stressful conditions can easily cause one to forget a critical step in taking a core or a CTD cast or launching a remotely operated vehicle (ROV). The start of each watch—0000, 0400, 0800, 1200, 1600, 2000—is an ideal time to review the upcoming operation, to receive a report of conditions during the previous

4 hours, and to relieve the watch on time. Thus as noted before, arriving for duty at least 10 minutes early, rested, and properly dressed, is expected conduct at sea.

Equipment

An exposition of all possible (and future) underwater instruments is not possible nor is it practicable in this, *The Oceanographer's Companion*. The magnetometer discussed earlier has the cable assembly molded to the 2-foot-long "fish." Other instruments such as a CTD will have a waterproof plug, and great care must be exercised as such plugs almost always have only one way that they can be connected. A 3-foot-long side scan sound navigation and ranging (SONAR) tow fish (Figure 14.1) is discussed next as it has both a detachable cable and a "weak link." A weak link can be found on anchors and for the same reason: if the tow fish or anchor gets caught in the bottom, the weak link breaks and the unit reverses and probably can be recovered.

Side scan sonar is an acoustic system for imaging the sea floor, and the tow fish is ordinarily towed fairly close to the bottom. The 600 kHz tow fish pictured in Figure 14.1 is a high-frequency unit easily capable of imaging a self-contained underwater breathing apparatus (SCUBA) diver or a small wreck. Water depth and towing vessel speed dictate how much side scan sonar tow fish cable is let out; each system has sets of tables giving the oceanographer safe operating variables. If the tow fish in Figure 14.1 snags on an obstruction and if the weak link breaks, the tow wire reverses and the tow fish is hauled backwards (i.e., the fins become the leading edge). Such an

FIGURE 14.1

Side scan sonar tow fish. The tow fish is hanging from the weak link. If the weak link breaks, the tow cable attachment point shifts to the eye at the rear near the stabilizing fins. The tow fish is about 3 feet long.

arrangement minimizes loss of the tow fish and the major portion of the investment in the system.

The electric cable connecting the tow fish with the shipboard instrumentation (often a computer with a special electronics board) must never be kinked. Figure 14.2 shows a "Chinese finger-grip" designed to spread the load of towing the fish. The finger-grip is nylon with Velcro to attach a tow cable to the underwater unit. The finger-grip spreads the strain of towing along a section of cable, about 2 feet in Figure 14.2, and prevents kinks. Note that the tow cable integrates towing with the signal cable as in the case of the magnetometer.

Moving from the sophisticated side scan sonar to a purely mechanical device older than the *Challenger* Expedition, consider the orange-peel grab in Figure 14.3. This grab, and variants such as the clam-shell grab, are designed to recover bottom samples in soft bottom environments. The orange-peel grab relies on sharp edges, but no springs as are seen in the Shipek grab sampler (a finger-eating device if ever there was one!). Most of these grabs are heavy and rely on gravity and/or springs to return an undisturbed bottom sample to the research vessel. The seagoing scientist or engineer using such devices is well advised to work alongside an experienced survey technician before attempting to "cock" the unit. Retaining all 10 fingers at the end of a seagoing career is more a badge of caution than a badge of honor.

Corers come in a wide variety of designs, but all are meant to return to the surface a history of oceanic sediment deposition over time. The simple gravity core (Figure 14.4) drives a weighted pipe into the sea bottom by shear force; the more sophisticated piston corer uses a release mechanism, gravity, and a vacuum producing piston to sink the pipe ever deeper into soft bottom. A piston core can be many tens of feet long and requires an unobstructed passageway to swing the unit aboard (an interesting challenge for naval architects).

Dredges are even less sophisticated sampling devices meant to return rocks and nodules to the seagoing geologist. Dredges can be as simple as a large (2- to 3-foot diameter) pipe that is dragged along the bottom, to box-mouthed units with chain bags to capture the hard materials. Manganese nodules varying in size from golf balls to newspaper-sized slabs have been recovered all over the seafloor. Dredging requires lowering the sampling device and dragging it along the bottom. The research vessel is well advised to rig a tension-measuring meter so as to minimize the risk of catastrophic failure in

FIGURE 14.2
Finger grip (black) wrapped around tow cable (gray). Two such finger grips would be used on the side scan sonar in Figure 14.1: one at the weak link and one on deck to attach the tow cable to a deck fixture such as a cleat.

FIGURE 14.3
Orange-peel grab for obtaining a bottom sample. The grab is lifted by the cable through the block after hitting bottom. The grab is about 2 feet tall and weighs 50 pounds.

the dredging cable. If the sampler is "hung up" on an outcrop, it is much safer to reverse course and dislodge the unit rather than risk snapping the cable.

Underwater cameras were developed for research purposes after World War II. "Doc Edgerton," the well-known "E" in EG&G, Inc., among others, developed strobes and shutter-less high-speed cameras that were lowered on the hydrographic wire and "flew" over the seafloor. Keeping the camera system an exact height above the bottom is accomplished using an acoustic "pinger." The pinger, about the size of a large loaf of bread, is mounted vertically so that one transducer points upward and the other downward; the difference in arrival of the direct ping and the ping reflected from the bottom is proportional to the height of the camera(s). One version of the EG&G camera system used two cameras and obtained stereographic photographs, some of which led to important geochemical discoveries such as manganese nodules and pavement.

Pingers often have 12-kHz-acoustic transducers so as to be compatible with research vessel operational echo sounders. Water depth "fathometers" for

FIGURE 14.4

Simple gravity core. Unit is about 3 feet tall without the core tube screwed in; typically the 3-inch-core tube adds another 2 feet to the overall length. The core weighs about 50 pounds and is lowered by a hydrographic cable from a winch. Often a release mechanism will free-fall the core 10–30 feet depending on the bottom type (sand, mud, etc.).

navigation and surveying are often designed for 12 kHz as signal attenuation is low. Sound speed ($c[z]$) in the ocean is a function of temperature, salinity, and pressure ($p = \rho g z$ from the hydrostatic equation), and can be determined directly by a sound velocimeter or from parametric equations. If precise water depth information is required, such as for nautical charting, then the

mean sounding velocity (MSV) must be obtained from $MSV(z) = \left[\dfrac{1}{z} \int_{0}^{z} \dfrac{dz}{c(z)} \right]^{-1}$

where z is the water depth. The equation for MSV(z) is the geometric mean of the sound speed profile ($c[z]$), and is solved iteratively. For pinger operations, the standard speed of sound, 820 fathoms/s (1500 m/s), leaves little error when lowering an instrument near to the seafloor.

Bathymetry, sometimes called fathometry or echo sounding, is the process of determining bottom topography (yes the term is often misused,

FIGURE 14.5
Swath bathymetry supplemented by side scan sonar. The multibeam system is shown sending and receiving an acoustic swath from the survey ship, and the towed side scan sonar is obtaining very high-resolution imagery of the seafloor.

but compare the etymology of "altimetry," "geometry," "calorimetry," "gravimetry," etc.). During the *HMS Challenger* Expedition (1872–1876), bathymetry was accomplished by lowering a sounding weight; during the *RV Meteor* Expedition (1925–1927) acoustics were used extensively for the first time; during the 1967 around-the-world trip of the NOAA Ship *Oceanographer*, vertically stabilized narrow beam echo sounding was state-of-the-art. Today, swath bathymetry (often coupled with side scan sonar—Figure 14.5) acoustically images the seafloor much as a weather satellite images Earth. Yet the fundamentals of "hydrographic surveying," nautical chart making, are unchanged: (1) positioning to determine geographic location, (2) time to account for tides and date, (3) sound speed to correct for MSV(z), and (4) recordkeeping to collate all the information.

Dive Operations

Herodotus, *ca.* 500 BC, documents the probable use of hollow reeds as snorkels, which is the earliest written account of diving (although carvings and pottery suggest even earlier dates). Aristotle, student of Plato and teacher of Alexander the Great (*ca.* 340 BC), wrote of diving bells. Leonardo da Vinci (1452–1519) drew air tanks and snorkels, and in 1650 Otto van Guericke invented the air pump. In 1715, Pierre Rémy de Beauve invented the diving "dress" that included a metal helmet and hoses, and six decades later another ingenious Frenchman (Fréminet *ca.* 1772) invented a diving dress using compressed air. In 1837, Augustus Siebe integrated a diving helmet with a closed "dress" to invent the forerunner of the modern diving suit—no longer was the diver constrained to a vertical position! Then, yes of course, in 1943 Emile Gagnan and Jacques-Yves Cousteau patent the dive pressure regulator so familiar to today's users of SCUBA.

Flashback: Flopping backwards overboard from the Zodiac inflatable boat, US Divers Aquamaster double hose regulator mouthpiece clenched between our teeth, into the very turbid Elizabeth River. No thoughts back two and a half millennia or farther. Just a safe inspection dive of the hull of the Ship *Discoverer*. Water really rough, dirty, dark..... Beginning to feel seasick! Call off the dive. Not all diving is fun.

Diving is an endeavor that can be the exhilarating moment of sitting on the seafloor, watching a school of fish swimming by followed by a large barracuda, to sampling bottom materials while studying sand waves off the mouth of the Delaware Bay, to going under the hull of a barnacle encrusted ship in waters requiring feel-only inspections for a leak. Few sport divers are trained at the level of a US Navy SEAL, but diving scientists and engineers should train to the level expected for underwater operations from UNOLS ships—that is at the level of the American Academy of Underwater Sciences. Organized in 1977, "The mission of the American Academy of Underwater Sciences (AAUS) is to facilitate the development of safe and productive scientific divers through education, research, advocacy, and the advancement of standards for scientific diving practices, certifications, and operations" (www.aaus.org).

The general procedure for a dive starts with filling out the dive plan, and posting it where a dive safety officer can inspect and approve it (or not). The person responsible for the field portion of the dive should be AAUS certified, meaning that he or she has taken the AAUS training course, and has logged 12 science dives in the past year. Information on the dive form will include the names and certificate number of the divers (three as a minimum with the third being the rescue diver who stays on the boat unless needed), the boat operator (who also must have a safe-boater certificate), and the cruise plan. Contact information for all personnel on the voyage is also mandatory as is assurance that the dive boat is properly equipped and has active communications. Most AAUS member organizations will have standard procedures to follow including knowing where the nearest decompression chamber is located.

When at a dive site, if the water is deep enough, it is usually best to place the anchor of the dive boat (see Chapter 10) up-wind or up-current from the underwater worksite. Pay out sufficient rode to bring the boat directly over the dive site, using at least a 5:1 ratio of rode to water depth. Stream a safety line off the stern using a float at the bitter end and have a throwable personal floatation device (PFD) at hand. Raise the anchor ball and the international signal flag "alpha," which must be kept unfurled with a stiffener. Set the dive safety watch and lower the dive ladder. State laws vary, but in general keep at least 300 feet away from other dive operations except if working in a navigation channel, where a lesser distance (usually 100 feet) is permissible. Maintain a sharp lookout to ensure that the anchor is not dragging, and watch for dangerous marine animals.

At the completion of the dive, after all the saltwater is rinsed from the SCUBA gear, and from the boat, and the motor is flushed, the divemaster must "close the loop" and inform the dive safety officer that the boat is returned and that all in the crew are safe. If an incident occurred during the dive, it must be fully documented and debriefed. As with any research vessel operation, SOLAS is the central concern.

Underwater Archeology

One of the most interesting and challenging dive operations is that of recovering artifacts that may be several millennia in age. Much of what has been learned about ships and shipping and naval warfare of past cultures comes from the exploration of underwater wrecks. Underwater archeology is not limited to shipwrecks, as for example, the destruction and submergence of Port Royal, Jamaica, from an earthquake and tsunami in 1692. Sunken cities, lost cargo, lost aircraft, burial sites, and inundated valleys converted to estuaries by sea level rise: all are examples where underwater archeology has revised understanding of human activities. The *Gilgamesh Epic*, perhaps humankind's oldest extant literature, set the stage for studying drowned villages in the Black Sea, now under 400 feet of water. It is not limited to seas, as lakes and rivers cover many interesting artifacts yet to be discovered.

Archeology is scientific discovery and preservation. Maritime archeology (a.k.a. marine archeology) is a branch of underwater archeology that is focused on ships and their cargo, routes, and history, whereas nautical archeology is a branch of maritime archeology that studies the ships themselves: design, materials, construction, and so forth. The terms are often used differently in different countries. Underwater archeology is distinct from underwater salvage. Salvage is recovering artifacts and returning them to the surface for commercial purposes.

In 2001, the United Nations Educational, Scientific and Cultural Organization (UNESCO) adopted a treaty titled the UNESCO Convention on the Protection of Underwater Cultural Heritage. Its purpose is to protect "all traces of human existence having a cultural, historical, or archeological character which have been partially or totally under water, periodically or continuously for at least 100 years." The Convention is related to UNCLOS, the United Nations Convention on the Law of the Sea (see Chapter 13); the United States is neither a part of UNCLOS nor the UNESCO Convention on the Protection of Underwater Cultural Heritage; most developed nations are not signatories either. Central to the issue under Admiralty Law is the definition of when the 100 years starts, as well as the definition of "cultural heritage." The Convention's underlying mission seems more to be that of preservation *in situ*, that is in place, leaving no opportunity for scientific recovery, enhanced academic knowledge, and public display.

Often the first indication of an underwater site of archeological interest is the report of local trawl fisherman hauling up items such as Roman-era amphorae from the seafloor. This is particularly true in the Mediterranean Sea, but has been documented as far away as Brazil by Sir Robert L. Marx (see Appendix A). While ROVs allow underwater archeology at great depths (such as that of the *SS Titanic*), much of the work is done by SCUBA divers (Figure 14.6) in shallow water. In waters 25 feet deep or less, a hookah may be a better air-supply system than compressed air in tanks; in deep water, the diver may need to adopt a mixed-gas system to minimize decompression and especially avoid nitrogen narcosis, or perhaps worse yet, the bends.

The first scientific task of archeology is mapping the site with as much accuracy as possible. This is particularly difficult underwater, but is easier now that the Global Positioning System (GPS) is so readily available and of low cost. Side scan sonar images are of great value, but quantitative measurements still require creating a baseline and using measuring tapes and trilateration to document the site particularly if the visibility is low. Using GPS-equipped surface buoys with mooring scope close to 1:1, which are anchored at the site, a baseline is established and trilateration can begin. Photography too is incredibly valuable, but it needs to be in the context of a well-mapped site. Buoys, survey markers, SCUBA, stereographic photogrammetry, sonar, GPS—all adding up to better underwater science.

FIGURE 14.6
Self-contained underwater breathing apparatus diver using an underwater metal detector to discover a stock anchor on the seafloor. (Photo courtesy of Sir Robert L. Marx, underwater archeologist.)

Underwater archeology of course is a complex subject requiring many years of training and field experience. Artifacts moved from water to air require complex chemical preparation to conserve them properly. Identification of objects requires in-depth expertise in art, ancient languages, and history. Legal rulings complicate ownership. Funding archeological expeditions is expensive and return on investment uncertain. Seagoing scientists and engineers are well advised to cautiously entertain commitment to such an expenditure of professional time. But then, follow your passion!

Additional Reading

Briggs, P. 1969. *Science Ship*. New York, NY: Simon & Schuster, 128 pp.

Exercises

1. You are assigned the task of taking two student divers to obtain seagrass samples from the flood shoal at a nearby inlet. The cruise takes place in daylight during spring break. Include in your write-up how much time to allot to the cruise and what time of day you would choose to get started (think weather). Trailering a 17' Boston Whaler, set up a cruise plan to carry the two divers to their sampling site and return. Who should know about the trip?
2. Take an online safe boating course. Most are national, but some states will have specific requirements (Florida for example requires all motorboat operators born after January 1, 1988 to have a certificate in their possession while operating the vessel).
3. Biological oceanographers might recognize this. What is it?

15

Celestial Navigation—Spherical Trigonometry, Spherical Triangles, Azimuth, Sextant Altitude, Amplitude, and Line of Position

Seagoing scientists and engineers depend on the professional mariners on larger oceanographic research vessels for marine positioning (the general term for determining the latitude and longitude of an observation). So, why would an oceanographer need to know about an old-fashioned method of finding one's location? Beyond the pure joy of learning, it gives us the perspective of how well older observations can be used: the finest Nansen bottle cast with ultraprecise reversing thermometers and titration-determined salinity is only as good as the place and time that it actually was observed. Navigation errors of ±10 nm or more are not uncommon in historical data sets.

Ancient cartographers were well aware that latitude could be determined from the sun and stars, and they also knew that longitude was a matter of knowing the correct time difference between two places. At sea, time at a reference longitude, say Greenwich, England, could be determined by the *lunar-distance method,* but it was cumbersome and often required mathematical skill beyond that of the ordinary seaman. Determining longitude at sea probably was the major scientific problem of the eighteenth century, so much so that the English Parliament offered the longitude prize of £20,000 in 1714. It was won by John Harrison, who invented the marine chronometer just prior to the time of the American Revolutionary War.

Spherical Trigonometry

Celestial navigation is based on spherical trigonometry. The sine law and cosine law of a spherical triangle have their parallels in plane trigonometry, except that the sum of angles in a spherical triangle can be between 180° and 540°. Consider a spherical triangle with angles A, B, and C, with opposite sides of a, b, and c. Sides a, b, and c are arcs of great circles, which are formed by the intersection of a plane passing through the sphere's center. The laws are as follows:

$$\cos a = \cos b \cos c + \sin b \sin c \cos A$$

and

$$\frac{\sin A}{\sin a} = \frac{\sin B}{\sin b} = \frac{\sin C}{\sin c}$$

An example of a spherical triangle (Figure 15.1) is formed by the three great circle arcs connecting the elevated north pole (P_n), the zenith (Z) of the Melbourne (Florida) observer ($\phi = 28°N$, $\lambda = 80°W$), and point G, the Greenwich (England) Observatory ($\phi = 51.5°N$, $\lambda = 0°$). The great circle arc (*a*) between the two cities would be

$$\cos a = \cos(90° - 28°)\cos(90° - 51.5°) +$$

$$\sin(90° - 28°)\sin(90° - 51.5°)\ \cos(80° - 0°)$$

In the preceding, $b = 90° - 28° = co - \phi$ of Melbourne, $c = 90° - 51.2° = co - \phi$ of Greenwich; $co - \phi$ is the colatitude. A bit of arithmetic gives $a = 62.428° = 1.0896$ rad, and recalling the geometric equation $S = r\theta$, where Earth's radius $r = 6378$ km and the *arc a* $= \theta$ is in radians, then the great circle distance is $S = 6949$ km or $S = 3752$ nm. It is useful when visualizing this triangle to plot the points discussed on a globe.

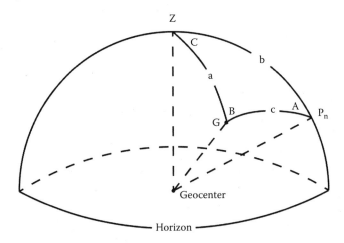

FIGURE 15.1
Spherical triangle with angles A, B, C and sides a, b, c. Point P_n is the North Pole, point Z is Melbourne, Florida's zenith, and point G is Greenwich, England. The horizon is a great circle everywhere 90° from Z. Sides a, b, c are arcs of great circles.

The navigation triangle is also a spherical triangle with the vertices being the elevated pole (North Pole [P_n] in the Melbourne–Greenwich example aforementioned), the point directly above the observer (the zenith, Z), and the geographic position (GP) of the celestial body (the GP is the point on Earth's surface intersecting an imaginary line from the geocenter to the celestial body, where the body's declination is the "celestial latitude," and the right ascension is the celestial longitude). The upper half of Figure 15.2 shows the basic geometry with the horizontal line being the observer's horizon, and Z directly overhead. The horizon is a great circle, but in Figure 15.2 it is a line due to the perspective view at zero elevation.

The angular elevation of the pole is the observer's latitude (i.e., *Polaris*, the pole star, is approximately 28° above the northern horizon at Melbourne, and 51.5° above the horizon at Greenwich, but the horizon is always the plane 90° from the zenith). The upper part of Figure 15.2 is called a meridian diagram because the semicircle from north to south is the arc of the great circle that passes through both the zenith (Z) and the elevated pole, and is the observer's meridian, or line of longitude. The circle below the meridian diagram is the view of the celestial sphere from the South Pole (P_s) circled by the equator, where M is the meridian of the observer. Meridians, which are great circles, appear as straight lines radiating out from the pole. The diagram in

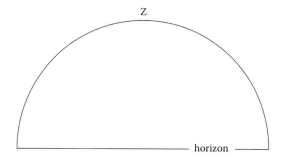

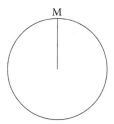

FIGURE 15.2
Upper panel is a meridian diagram with the zenith (Z) being directly overhead and the horizon a plane at right angles. The lower panel is a time diagram, which views Earth from above the South Pole (P_s) with the local meridian (M) radiating out from P_s.

the lower part of Figure 15.2 is called a time diagram, and its use in celestial navigation is that of a "longitude clock."

In this much abbreviated venture into celestial navigation, the path of the sun from sunrise to noon will be considered. The first example will be sunrise, followed by a mid-morning sunsight, and finally the sun's position at local apparent noon (LAN).

Morning Amplitude

Due to atmospheric refraction, the center of the sun (symbol $\odot$) at sunrise (when it is exactly on the celestial horizon) is when the lower limb of the sun appears to be about a solar semi-diameter above the visible horizon. The zenith distance at that moment, that is the angular distance from Z to $\odot$, is 90°. The direction to the sun, its azimuth, is the angle from true north toward the celestial body; many navigators will check their gyrocompass and/or their magnetic compass at this time of morning or evening because the azimuth (Z_n) is easy to calculate. This special azimuth at sunrise or sunset is called the sun's amplitude (A). The geometry is given in Figure 15.3.

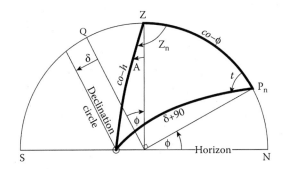

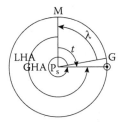

FIGURE 15.3

At sunrise or sunset the coaltitude is 90°. In this example the solar declination is 10°S, and the observer's latitude is 28°N.

In Figure 15.3, the time of the year is winter, and the sun's declination (δ) is drawn to be 10°S (the declination circle is a small circle of the celestial sphere parallel to Q the celestial equator). The zenith distance is labeled co-h because the symbol for the angular altitude of the sun above the horizon is "*h*," thus the zenith distance is also referred to as the coaltitude ($co - h$). The latitude and longitude of the observer in this example is (again) Melbourne, Florida ($\phi = 28°N$, $\lambda = 80°W$); thus the colatitude $(co - \phi)$ is 90°–28° = 62°. Note that the elevated pole (P_n) is 28° above the northern horizon. Since the declination is 10°S, the codeclination $(co - \delta)$ is 90° + 10° = 100° (if the sun's declination were north, then $co-\delta$ would be 90°–δ). For the scenario in Figure 15.3, the sides of the navigational triangle are $co-h = 90°$, $co-\phi = 62°$, and $co-\delta = 100°$, and to determine Z_n or A, the law of cosines is required. The time diagram in the lower half of Figure 15.3 is drawn to illustrate the hour angles (HA) at this time of day.

In Figure 15.3, the longitude (λ) of Melbourne is 80°W; therefore, the Greenwich meridian (G) must be 80° east of M (which is the local meridian of the observer). The sun (O) from the perspective of the south geographic pole (P_s) is not yet overhead at Greenwich; the Greenwich Hour Angle (GHA) is measured from the upper branch of the Greenwich meridian (G) westward (counterclockwise) to the sun. The local hour angle (LHA) is measured westward (counterclockwise) from the upper branch of the local meridian (M). The meridian angle (*t*) is measured eastward (clockwise) or westward (counterclockwise) from M; t is eastward at sunrise.

From the law of cosines for this scenario

$$\cos(\delta + 90°) = \cos(co - \phi)\cos(co - h) + \sin(co - \phi)\sin(co - h)\cos(Z_n)$$

or

$$\cos(\delta + 90°) = \sin(\phi)\sin(h) + \cos(\phi)\cos(h)\cos(Z_n)$$

Recalling the trigonometric functions of sums of angles

$$\cos(x \pm y) = \cos x \cos y \pm \sin x \sin y$$

and

$$\sin(x \pm y) = \sin x \cos y \pm \cos x \sin y$$

$\cos(\delta + 90°) = -\sin(\delta)$, and $h = 0°$, so $-\sin(\delta) = \cos(\phi)\cos(Z_n)$, or $\cos(Z_n) = -\sin(\delta)\sec(\phi)$. The amplitude is $A = Zn - 90°$, and from the trigonometric identity $\sin(Z_n - 90°) = \sin(Z_n)\cos(90°) - \cos(Z_n)\sin(90°)$, or $\sin(A) = -\cos(Z_n)$ the equation for amplitude (A) is

$$\sin(A) = \sin(\delta)\sec(\phi)$$

Numerically, since $\delta = 10°S$ and $\phi = 28°N$, the amplitude at sunrise is $\sin(A) = \sin(10°)\sec(28°)$, or $A = S11.3°E$ (the prefix is S because the declination is south, and the suffix is E because the sun is rising). Alternately, $\cos(Z_n) = -\sin(10°)\sec(28°)$, or $Z_n = 101.3°$ true.

Morning Sunline

In the lower half of Figure 15.3, it is seen that the sun moves westward (counterclockwise from the perspective of the South Pole, P_s). From the law of sines for a spherical triangle

$$\frac{\sin t}{\sin(co - h)} = \frac{\sin Z_n}{\sin(\delta + 90°)}$$

the meridian angle t is calculated to be $t = 84.7°E$, and the morning sun lies $84.7° - 80° = 4.7°$ east of G, the Greenwich meridian, as can be seen in the time diagram in the lower portion of Figure 15.3. The GHA of the sun is then $360° - 4.7° = 355.3°$, and the LHA is $355.3° - 80° = 275.3°$.

The next example is following the sun's path from sunrise toward noon, a mid-morning event, again assuming that the declination is $\delta = 10°S$, and the observer is at Melbourne, Florida ($\phi = 28°N$, $\lambda = 80°W$). Figure 15.4 summarizes the celestial triangle on a meridian diagram and the time diagram (again) for mid-February.

At mid-morning the solar altitude (h) is shown to be (Figure 15.4) some degrees above the southeast horizon. The Z_n is larger than at sunrise because the sun is rising toward its maximum altitude at LAN when it will be due south ($Z_n = 180°$). The colatitude ($co - \phi$) and codeclination ($co - \delta$) are the same as shown in Figure 15.3, but the coaltitude ($co - h$) now has decreased from 90°. In the time diagram (lower part of Figure 15.4), the sun ($\odot$) is seen to have moved from east of Greenwich (G) toward the local meridian (M); the longitude (λ) is 80°W, as before, but the meridian angle (t) has decreased; the GHA of the sun and the LHA have both changed due to the apparent solar motion. Knowing the exact time at Greenwich, commonly known as Greenwich Mean Time (GMT), allows the navigator to look up the GHA in the *Nautical Almanac*. Thus, the GP of the sun (or any celestial body at any given moment in time) can be plotted using the GHA for longitude (corrected for east or west) and δ for latitude.

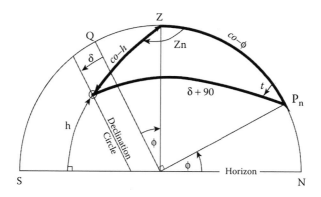

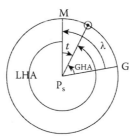

FIGURE 15.4
Sun at mid-morning using the same declination and latitude as in Figure 15.3.

In the exercise that follows, the solar altitude and azimuth for a meridian angle of $t = 11°$ (chosen as an example only) is calculated and compared with that in Hydrographic Office Publication No. 229, *Sight Reduction Tables for Marine Navigation*, a standard set of tables used by mariners. From the law of cosines for the spherical triangle in Figure 15.4

$$\cos(co - h) = \cos(co - \phi)\cos(\delta + 90°) + \sin(co - \phi)\sin(\delta + 90°)\cos(t)$$

Then

$$\sin(h) = \sin(\phi)\cos(\delta + 90°) + \cos(\phi)\sin(\delta + 90°)\cos(t)$$

But, from the functions of sums of angles

$$\cos(\delta + 90°) = \cos(\delta)\cos(90°) - \sin(\delta)\sin(90°) = -\sin(\delta)$$

and

$$\sin(\delta + 90°) = \sin(\delta)\cos(90°) + \cos(\delta)\sin(90°) = \cos(\delta)$$

Accordingly

$$\sin(h) = -\sin(\phi)\sin(\delta) + \cos(\phi)\cos(\delta)\cos(t)$$

For another example for Melbourne ($\phi = 28°N$, $\lambda = 80°W$), let $\delta = 10°S$, $t = 11°E$.
Then

$$\sin(h) = -\sin(28°)\sin(10°) + \cos(28°)\cos(10°)\cos(11°)$$

$$h = 50°32.2'$$

If the sun's declination were 10°N instead of 10°S, the altitude (h) would be

$$\cos(co - h) = \cos(co - \phi)\cos(co - \delta) + \sin(co - \phi)\sin(co - \delta)\cos(t)$$

or

$$\sin(h) = \sin(\phi)\sin(\delta) + \cos(\phi)\cos(\delta)\cos(t)$$

and for the example aforesaid, $h = 69°14.5'$. So in summer, when the declination is north, the sun is higher in the sky at the same moment in time, as expected. Similarly, to calculate the azimuth for $\delta = 10°S$, using the law of cosines

$$\cos(100°) = \sin(50.54°)\cos(62°) + \cos(50.54°)\sin(62°)\cos(Z_n)$$

$$Z_n = 162.8° \text{ true}$$

Figure 15.5 is a copy of page 207 from H.O. Pub. No. 229, Vol. 2. The top of the table is for latitude contrary to declination, which is the case herein ($\phi = 28°N$, $\delta = 10°S$); the bottom is for latitude and declination of the same name. For $t = 11°E$ the LHA $= 360° - 11° = 349°$. Entering the sixth column from the left for $\phi = 28°$, and scanning down the column labeled "Dec." (i.e., δ) to 10°, the value calculated earlier is seen to be identical with the tabulated one; similarly the azimuth (Z_n in column six) is also exactly as calculated earlier. H.O. Pub No. 229 (and its predecessor H.O. Pub No. 214) allow the mariner to determine position without using a calculator—a handy capability when the batteries fade!

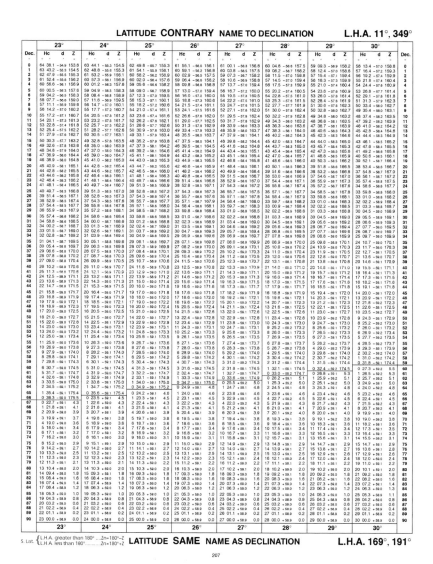

FIGURE 15.5

Page 207 from US Navy Hydrographic Office Publication No. 229 for the example in Figure 15.4.

Local Apparent Noon

As Earth rotates the sun appears to move farther south until at LAN, it reaches its maximum altitude for the day. At this point in time, the celestial triangle becomes an arc along the meridian diagram, as shown in Figure 15.6. The mathematics become much simpler because $Zn = 180°$:

$$\cos(\delta + 90°) = \sin(\phi)\sin(h) + \cos(\phi)\cos(h)\cos(Z_n)$$

becomes

$$\cos(\delta + 90°) = \sin(\phi)\sin(h) - \cos(\phi)\cos(h) = -\cos(\phi + h)$$

But, from the identity

$$\cos x = -\cos(180° - x)$$

$$-\cos(\delta + 90°) = \cos(180° - \delta - 90°) = \cos(90° - \delta)$$

or

$$\cos(90° - \delta) = \cos(\phi + h), \text{ or } 90° - \delta = \phi + h$$

Finally

$$h = 90° - 10° - 28° = 52°$$

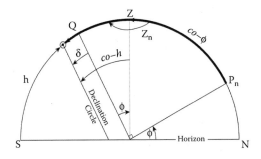

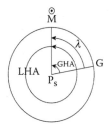

FIGURE 15.6
Sun at local apparent noon for the same declination and latitude used in Figures 15.3 and 15.4.

Of course the calculation of h in this example can be done by direct inspection of Figure 15.6. The general navigation problem is to determine latitude from the observation of the sun at LAN, and to a lesser extent the longitude as well, if the exact GMT of LAN can be observed. The GHA (see the lower panel of Figure 15.6) at LAN is exactly equal to the observer's longitude, that is, at LAN, λ = GHA. If a series of sun sights of h are started before LAN and continued afterward, the time series of altitudes will form an arc with LAN at the peak. The navigator then consults the *Nautical Almanac*, determines GHA in degrees for the GMT of LAN, and determines the longitude exactly. Thus, a fix is possible at local noon if the time series of altitudes is temporally refined enough.

Morning Star Sight

Now to illustrate the navigational triangle with a morning star sight on Vega. The date (again) is February 22, 2000 (Figure 15.7); the navigational triangle is illustrated in Figure 15.8. From the *Nautical Almanac* (page not shown) Vega's declination is about 39°N, and its position relative to the First Point of Aries (Υ), known as the *sidereal hour angle (SHA*)*, is about 81°. The First Point of Aires is the location of the *Vernal Equinox*—that point on the celestial equator where the sun crosses from south declination to north declination moving along the ecliptic. Motion of the First Point of Aires is known as *sidereal time*, that is, the hour angle of the vernal equinox, symbolized by Υ, as it rotates throughout the day. Earth's precession is about $\dfrac{360°}{26,000 \text{ years}} = 0.01°$ per year, so that the declination and SHA of stars changes very slowly.

In this example, morning nautical twilight (February 22, 2000) in the Nautical Almanac is 05^{h}42^m Local Mean Time (LMT) (75th meridian time zone), see Figure 15.7. However, Earth rotates 15° per hour, so Melbourne (ϕ = 28°N, λ = 80°W) is $\left(80° - 75°\right) \times \dfrac{15°}{\text{hour}} = 20$ minutes west of the standard meridian at 75°W; therefore, nautical twilight at Melbourne is at 05^{h}42^m + 20^m = 06^{h}02^m LMT or 11^{h}02^m GMT. From the Nautical Almanac (page not shown), the GHA of Aires (GHAΥ) is abstracted as follows:

GMT	GHAΥ
1100	316° 40.2′
1102	
1200	331° 42.7′

2000 FEBRUARY 21, 22, 23 (MON., TUES., WED.)   45

SUN / MOON

UT (d h)	SUN GHA	SUN Dec	MOON GHA	v	MOON Dec	d	HP
21 00	176 33.9	S10 53.1	338 58.3	10.4	N 7 49.1	11.6	58.6
01	191 33.9	52.2	353 27.7	10.4	7 37.5	11.7	58.5
02	206 34.0	51.3	7 57.1	10.6	7 25.8	11.7	58.5
03	221 34.1 ..	50.4	22 26.7	10.6	7 14.1	11.7	58.5
04	236 34.2	49.5	36 56.3	10.7	7 02.4	11.8	58.4
05	251 34.2	48.6	51 26.0	10.7	6 50.6	11.7	58.4
06	266 34.3	S10 47.7	65 55.7	10.8	N 6 38.9	11.8	58.4
07	281 34.4	46.8	80 25.5	10.9	6 27.1	11.9	58.3
08	296 34.4	45.9	94 55.4	10.9	6 15.2	11.8	58.3
M 09	311 34.5 ..	45.0	109 25.3	11.0	6 03.4	11.9	58.3
O 10	326 34.6	44.1	123 55.3	11.1	5 51.5	11.8	58.2
N 11	341 34.6	43.2	138 25.4	11.1	5 39.7	11.9	58.2
D 12	356 34.7	S10 42.3	152 55.5	11.1	N 5 27.8	11.9	58.2
A 13	11 34.8	41.4	167 25.6	11.3	5 15.9	12.0	58.1
Y 14	26 34.9	40.5	181 55.9	11.3	5 03.9	11.9	58.1
15	41 34.9 ..	39.6	196 26.2	11.3	4 52.0	11.9	58.1
16	56 35.0	38.7	210 56.5	11.4	4 40.1	12.0	58.0
17	71 35.1	37.8	225 26.9	11.5	4 28.1	12.0	58.0
18	86 35.1	S10 36.9	239 57.4	11.5	N 4 16.1	11.9	58.0
19	101 35.2	36.0	254 27.9	11.6	4 04.2	12.0	57.9
20	116 35.3	35.1	268 58.5	11.6	3 52.2	12.0	57.9
21	131 35.4 ..	34.2	283 29.1	11.7	3 40.2	12.0	57.9
22	146 35.4	33.3	297 59.8	11.7	3 28.2	12.0	57.8
23	161 35.5	32.4	312 30.5	11.8	3 16.2	12.0	57.8
22 00	176 35.6	S10 31.5	327 01.3	11.8	N 3 04.2	11.9	57.8
01	191 35.7	30.6	341 32.1	11.9	2 52.3	12.0	57.7
02	206 35.7	29.6	356 03.0	12.0	2 40.3	12.0	57.7
03	221 35.8 ..	28.7	10 34.0	11.9	2 28.3	12.0	57.7
04	236 35.9	27.8	25 04.9	12.1	2 16.3	12.0	57.6
05	251 36.0	26.9	39 36.0	12.1	2 04.3	12.0	57.6
06	266 36.0	S10 26.0	54 07.1	12.1	N 1 52.3	12.0	57.6
07	281 36.1	25.1	68 38.2	12.2	1 40.3	11.9	57.5
08	296 36.2	24.2	83 09.4	12.2	1 28.4	12.0	57.5
T 09	311 36.3 ..	23.3	97 40.6	12.3	1 16.4	11.9	57.5
U 10	326 36.3	22.4	112 11.9	12.3	1 04.5	12.0	57.4
E 11	341 36.4	21.5	126 43.2	12.3	0 52.5	11.9	57.4
S D 12	356 36.5	S10 20.6	141 14.5	12.4	N 0 40.6	11.9	57.4
A 13	11 36.6	19.7	155 45.9	12.4	0 28.7	11.9	57.3
Y 14	26 36.7	18.8	170 17.3	12.5	0 16.8	11.9	57.3
15	41 36.7 ..	17.9	184 48.8	12.5	N 0 04.9	11.9	57.3
16	56 36.8	16.9	199 20.3	12.6	S 0 07.0	11.9	57.2
17	71 36.9	16.0	213 51.9	12.6	0 18.9	11.8	57.2
18	86 37.0	S10 15.1	228 23.5	12.6	S 0 30.7	11.9	57.2
19	101 37.1	14.2	242 55.1	12.7	0 42.6	11.8	57.1
20	116 37.1	13.3	257 26.8	12.7	0 54.4	11.8	57.1
21	131 37.2 ..	12.4	271 58.5	12.7	1 06.2	11.7	57.1
22	146 37.3	11.5	286 30.2	12.8	1 17.9	11.8	57.0
23	161 37.4	10.6	301 02.0	12.8	1 29.7	11.7	57.0
23 00	176 37.5	S10 09.7	315 33.8	12.8	S 1 41.4	11.7	57.0
01	191 37.5	08.8	330 05.6	12.9	1 53.1	11.7	56.9
02	206 37.6	07.8	344 37.5	12.9	2 04.8	11.7	56.9
03	221 37.7 ..	06.9	359 09.4	13.0	2 16.5	11.6	56.9
04	236 37.8	06.0	13 41.4	12.9	2 28.1	11.6	56.8
05	251 37.9	05.1	28 13.3	13.0	2 39.7	11.6	56.8
06	266 37.9	S10 04.2	42 45.3	13.0	S 2 51.3	11.5	56.8
W 07	281 38.0	03.3	57 17.3	13.1	3 02.8	11.6	56.7
E 08	296 38.1	02.4	71 49.4	13.1	3 14.4	11.4	56.7
D 09	311 38.2 ..	01.4	86 21.5	13.1	3 25.8	11.5	56.7
N 10	326 38.3	10 00.5	100 53.6	13.1	3 37.3	11.4	56.6
E 11	341 38.4	9 59.6	115 25.7	13.1	3 48.7	11.4	56.6
S D 12	356 38.4	S 9 58.7	129 57.8	13.2	S 4 00.1	11.4	56.6
A 13	11 38.5	57.8	144 30.0	13.2	4 11.5	11.3	56.5
Y 14	26 38.6	56.9	159 02.2	13.2	4 22.8	11.3	56.5
15	41 38.7 ..	56.0	173 34.4	13.3	4 34.1	11.3	56.5
16	56 38.8	55.0	188 06.7	13.2	4 45.4	11.2	56.4
17	71 38.9	54.1	202 38.9	13.3	4 56.6	11.2	56.4
18	86 39.0	S 9 53.2	217 11.2	13.3	S 5 07.8	11.2	56.4
19	101 39.0	52.3	231 43.5	13.4	5 19.0	11.1	56.3
20	116 39.1	51.4	246 15.9	13.3	5 30.1	11.1	56.3
21	131 39.2 ..	50.5	260 48.2	13.4	5 41.2	11.0	56.3
22	146 39.3	49.5	275 20.6	13.3	5 52.2	11.0	56.2
23	161 39.4	48.6	289 52.9	13.4	S 6 03.2	11.0	56.2
SD 16.2 d 0.9			SD 15.9		15.6		15.4

Twilight / Sunrise / Moonrise

Lat.	Naut.	Civil	Sunrise	21	22	23	24
N 72	05 47	07 05	08 18	19 14	21 08	22 58	24 49
N 70	05 48	06 58	08 03	19 20	21 06	22 49	24 31
68	05 49	06 53	07 52	19 25	21 05	22 41	24 17
66	05 49	06 49	07 42	19 28	21 03	22 35	24 05
64	05 50	06 45	07 34	19 32	21 02	22 30	23 55
62	05 50	06 41	07 27	19 35	21 02	22 26	23 47
60	05 50	06 38	07 21	19 37	21 01	22 22	23 40
N 58	05 50	06 35	07 15	19 39	21 00	22 18	23 34
56	05 49	06 33	07 11	19 41	21 00	22 15	23 29
54	05 49	06 30	07 06	19 43	20 59	22 13	23 24
52	05 49	06 28	07 02	19 44	20 58	22 10	23 19
50	05 48	06 26	06 59	19 46	20 58	22 08	23 15
45	05 47	06 21	06 51	19 49	20 57	22 03	23 07
N 40	05 46	06 17	06 45	19 51	20 56	21 59	23 00
35	05 44	06 13	06 39	19 54	20 56	21 56	22 54
30	05 42	06 10	06 34	19 56	20 55	21 52	22 48
20	05 37	06 03	06 25	19 59	20 54	21 47	22 39
N 10	05 32	05 56	06 18	20 02	20 53	21 43	22 31
0	05 25	05 49	06 10	20 05	20 52	21 38	22 24
S 10	05 16	05 41	06 03	20 08	20 52	21 34	22 16
20	05 06	05 32	05 55	20 11	20 51	21 30	22 09
30	04 52	05 21	05 45	20 14	20 50	21 25	22 00
35	04 42	05 14	05 40	20 16	20 49	21 22	21 55
40	04 32	05 05	05 34	20 18	20 49	21 19	21 49
45	04 18	04 55	05 26	20 21	20 48	21 15	21 42
S 50	04 00	04 43	05 17	20 24	20 47	21 10	21 34
52	03 52	04 37	05 13	20 25	20 47	21 08	21 30
54	03 42	04 31	05 09	20 27	20 47	21 06	21 26
56	03 31	04 23	05 04	20 28	20 46	21 04	21 22
58	03 18	04 15	04 58	20 30	20 46	21 01	21 17
S 60	03 02	04 05	04 52	20 32	20 45	20 58	21 11

Sunset / Twilight / Moonset

Lat.	Sunset	Civil	Naut.	21	22	23	24
N 72	16 11	17 24	18 43	08 57	08 43	08 31	08 17
N 70	16 25	17 30	18 41	08 48	08 42	08 35	08 29
68	16 37	17 36	18 40	08 42	08 41	08 39	08 38
66	16 47	17 40	18 40	08 36	08 39	08 42	08 45
64	16 55	17 44	18 39	08 31	08 38	08 45	08 52
62	17 02	17 47	18 39	08 27	08 38	08 48	08 58
60	17 08	17 50	18 39	08 23	08 37	08 50	09 03
N 58	17 13	17 53	18 39	08 20	08 36	08 52	09 07
56	17 18	17 56	18 39	08 17	08 36	08 53	09 11
54	17 22	17 58	18 39	08 14	08 35	08 55	09 15
52	17 26	18 00	18 39	08 12	08 35	08 56	09 18
50	17 29	18 02	18 40	08 10	08 34	08 57	09 21
45	17 37	18 07	18 41	08 05	08 33	09 00	09 27
N 40	17 43	18 11	18 42	08 01	08 32	09 02	09 32
35	17 49	18 14	18 44	07 57	08 32	09 04	09 37
30	17 53	18 18	18 46	07 54	08 31	09 06	09 41
20	18 02	18 24	18 50	07 49	08 30	09 09	09 48
N 10	18 10	18 31	18 56	07 44	08 29	09 12	09 54
0	18 17	18 38	19 02	07 39	08 28	09 15	10 00
S 10	18 24	18 46	19 11	07 34	08 27	09 17	10 06
20	18 32	18 55	19 21	07 29	08 26	09 20	10 12
30	18 41	19 06	19 35	07 24	08 24	09 23	10 20
35	18 47	19 13	19 44	07 20	08 24	09 25	10 24
40	18 53	19 21	19 55	07 17	08 23	09 27	10 28
45	19 00	19 31	20 08	07 12	08 23	09 29	10 33
S 50	19 09	19 43	20 25	07 07	08 21	09 32	10 40
52	19 13	19 49	20 34	07 04	08 20	09 33	10 44
54	19 17	19 55	20 43	07 02	08 19	09 34	10 47
56	19 22	20 02	20 54	06 59	08 19	09 36	10 51
58	19 27	20 11	21 07	06 55	08 18	09 38	10 55
S 60	19 34	20 20	21 22	06 51	08 17	09 40	10 59

SUN / MOON

Day	Eqn. of Time 00h	Eqn. of Time 12h	Mer. Pass.	Mer. Pass. Upper	Mer. Pass. Lower	Age	Phase
21	13 48	13 41	12 14	01 27	13 52	16	96
22	13 36	13 34	12 14	02 16	14 40	17	90
23	13 30	13 26	12 13	03 03	15 27	18	83

FIGURE 15.7

Page 45 from the Nautical Almanac for February 21–23, 2000.

Using linear interpolation, GHAϒ (at 1102 GMT) = 317°10.3′.

The relationships between Vega's *SHA*∗, the Greenwich Hour Angle of Aires (*GHAϒ*), the LHA of Vega (*LHA*∗), and the meridian angle (*t*), are shown in the time diagram (lower half of Figure 15.8). The general equation is $LHA* = GHA\Upsilon + SHA* -\lambda$. In this example, the meridian angle *t* is east because $LHA* > 180°$.

A morning star sight is taken on Vega (Earth's future pole star in the year 13,727) at nautical twilight, that is, 1102 GMT. The observed altitude (h_o) from the navigator's corrected sextant altitude will be compared to the calculated altitude (h_c) using the assumed position (AP) of Melbourne as before. In this example (again, chosen purely for the sake of illustrating the process), the observed altitude is $h_o = 53°42.6′$. From the *Nautical Almanac*, Vega's $SHA* = 80°46.2′$; its declination $\delta = 38°46.8′N$. Since declination is the same name as latitude, the spherical triangle is as follows:

$$\cos(co - h) = \cos(co - \delta)\cos(co - \phi) + \sin(co - \delta)\sin(co - \phi)\cos(t)$$

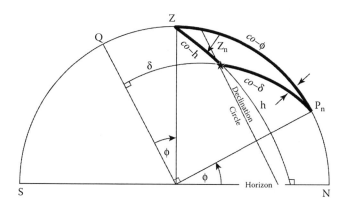

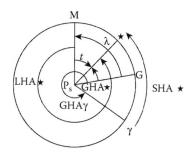

FIGURE 15.8

Example of morning star sight on Vega from Melbourne ($\phi = 28°N$) at nautical twilight.

or

$$\sin(h) = \sin(\delta)\sin(\phi) + \cos(\delta)\cos(\phi)\cos(t)$$

The meridian angle is given by $360° - t = GHAY + SHA* - \lambda = LHA*$ (see time diagram, Figure 15.8): $t = -(317°10.3' + 80°46.2' - 80°00.0') + 360° = 42°03.5'$ E. So, the computed altitude (h_c)

$$\sin(h_c) = \sin(38°46.8')\sin(28°) + \cos(38°46.8')\cos(28°)\cos(42°3.5')$$

$$\therefore h_c = 53°37.0'$$

The computed Z_n of Vega at this time of the morning from the sine law:

$$\sin(Z_n) = \frac{\sin(t)\cos(\delta)}{\cos(h)} = \frac{\sin(42.06°)\cos(38.78°)}{\cos(53.62°)} = 0.8805$$

$$Z_n = 61°41.8' \text{true}$$

Using declination and GHA, the GP (Figure 15.9) of Vega at this instant of time would be $\phi = 38°46.8'$N, and $\lambda = 317°10.3' + 80°46.2' - 360° = 37°56.5'$W.

In the routine practice of celestial navigation, the ship's navigator chooses an AP for ease of entering the *Sight Reduction Tables* (see Figure 15.5). With calculations as in these examples, the navigator would most likely choose the *dead reckoning* position as the AP (see Chapter 6). Consider then that Melbourne is assumed to be at $\phi = 28°$N, $\lambda = 80°$W, and the computed altitude of Vega is $h_c = 53°37.0'$. To complete the example shown in Figure 15.9, the observed corrected sextant altitude is $h_o = 53°42.6'$ as stated earlier. Note that the computed coaltitude $co - h_c = 90° - 53°37.0' = 36°23.0'$ and the observed coaltitude $co - h_o = 90° - 53°42.6' = 36°13.4'$. Thus, the observer is $36°23.0' - 36°13.4' = 5.6'$ closer to Vega's GP than assumed in the AP.

The line of position (LOP) also is $53°42.6' - 53°37.0' = 5.6' = 5.6$ nm toward $061.8°$ true from the AP of $\phi = 28°$N, $\lambda = 80°$W in Melbourne, a distance also called the *altitude intercept* (Figure 15.9). Navigators use the mnemonic CGA to remember that if the **C**omputed altitude is **G**reater than the observed altitude, then the observer's position is **A**way from the AP (**C**omputed **G**reater **A**way). Using a universal plotting sheet, the LOP can now be plotted from the AP toward the GP of Vega. Figure 15.9 illustrates the relationships. While the navigator's position is anywhere along the circle of equal observed altitude (h_o), for short distances on a Mercator universal plotting sheet or nautical chart, it can be approximated by a straight line and is labeled LOP. Note that a circle of equal altitude is a small circle of the celestial sphere orthogonal to the line from the geocenter to the GP, whereas $co - \delta$, $co - h_o$, and $co - \phi$ are arcs of great circles as noted in Figure 15.1.

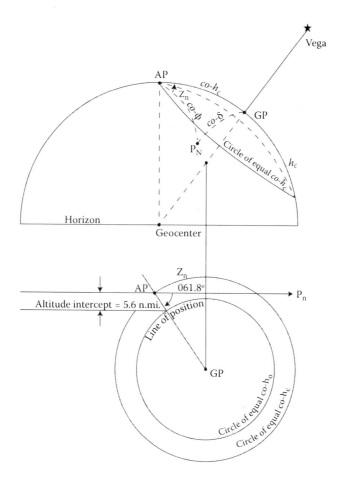

FIGURE 15.9
Example of a morning star sight on Vega from Melbourne. The upper panel shows the configuration of the assumed position (AP) and the geographic position of Vega (GP) along the great circle passing through AP and GP; the North Pole and the navigational triangle is behind the view. The lower panel is a plan view of the line of position (LOP).

In practice, celestial navigation is an art. The sextant must be adjusted properly: index mirror perpendicular to the frame of the sextant, horizon glass perpendicular to the frame, and index error removed (sextant at 0°). In addition, navigators will determine their own "personal error" (a personal error can be determined from the triangle that forms when plotting three LOPs; if closing the triangle into a point can be accomplished by adjusting each LOP the same amount, that amount is the personal error). Sextant altitudes also need to be corrected for atmospheric refraction and height of eye of the observer (both corrections contained in the *Nautical Almanac*). Forms are available to record all the observations and corrections, and to remind the navigator to apply them all properly. Finally, the chronometer error must be determined by comparing the clock with a time signal such as from the National Institute of Standards and Technology radio station WWV on 5, 10, 15, and 20 MHz (several frequencies are simultaneously broadcast to ensure reception). The ship's navigator (usually the Second Mate) will maintain records of the chronometer rate as well. WWV is the equivalent of having a "time ball" available 24/7, a nice improvement over Robert Wauchope's invention in 1829 (Chapter 1).

Additional Reading

Culter, T.J. 2004. *Dutton's Nautical Navigation*, 15th Edition. Annapolis, MD: U.S. Naval Institute, 464 pp.

Exercises

1. Herodotus (484–425 BC) wrote that the Phoenicians circumnavigated Africa *ca.* 600 BC, which was judged credible as they reported sighting the noonday sun in the north. Assuming that the Phoenicians rounded the Cape of Good Hope on June 21st, what latitude did they measure?

2. You observe an afternoon sun-line in Melbourne: $\phi = 28°N$, $\lambda = 80°W$ (AP) on April 15, 2000 at $16^h07.9^m$ EDT; the observed sextant altitude $h_o = 46°48.9'$. Look up the solar data in the *Nautical Almanac* (http://navsoft.com/downloads.html); calculate the GMT, GHA☉, LHA☉, δ, and t of the sun-line; draw the meridian diagram looking from the west; sketch the triangle: *co*-δ, *co*-ϕ, *co*-*h*; draw the time diagram looking from P_s; sketch M, G, λ, t, LHA☉, GHA☉; calculate the solar altitude and azimuth; and plot the AP, the sun's azimuth (Z_n), and the LOP on a universal plotting sheet.

3. This instrument might have adorned the master's cabin in the nineteenth century. What is this?

Appendix A: Timeline of Important Events in Ocean Science and Engineering

Date	Person	Event
BC 2300	Sargon of Akkad	Oldest known map (Babylon); flat Earth with a saltwater river encircling the land
BC 1500	Egyptians	Gnomon, predecessor to the sundial invented; world's oldest scientific instrument
BC 850	Homer	World conceived as land adjacent to Mediterranean Sea surrounded by Oceanus, the land-circling waters beyond
BC 600	Herodotus	Reported that the Phoenicians circumnavigated Africa
BC 585	Thales of Miletus	Gnomonic chart projection invented; brought navigational astronomy to Greeks from the Phoenicians; Earth a sphere
BC 550	Anaximander	Introduced gnomon to Greeks; world map—land surrounded by Oceanus (later printed by Hecataeus, *ca.* 500 BC)
BC 525	Pythagoras	School of philosophy founded in Italy; Pythagorean Theorem; Earth a sphere
BC 386	Eudoxus	First globe (2-m diameter)
BC 330	Aristotle	Published *Meteorologica*; scientific method established
BC 250	Aristarchus	Heliocentric theory
	Pollonius of Perga	Astrolabe invented
BC 240	Eratosthenes	Estimated Earth's circumference at about 25,200 statute miles
BC 220	Archimedes	Formulated the laws of hydrostatics
BC 150	Hipparchus	Applied astronomic techniques to mapping; invented stereographic and orthographic chart projections; used astrolabe to determine latitude; used latitude and longitude to define geographic positions; proposed using solar eclipses to determine longitude
BC 100	Posidonius	Reports a sounding of more than 1000 fathoms
	Posidonius	Calculated Earth's circumference at 24,000 statute miles
BC 57	Romans	*Mille Passum* 1000 double paces = one mile
BC 50	Strabo	Wrote his *Geography* and described climatic zones; calculated Earth's circumference to be 18,000 statute miles
AD 150	Claudius Ptolemy	Described a factual map of the known world in his *Geography*; conic chart projection published in *Cosmographia*
AD 391	Aurelius	Great Library of Alexandria destroyed

(Continued)

(Continued)

Date	Person	Event
AD 851	Sulaiman el-Tagir	Kamal in use by Arabs to measure latitude; persistent tradition that Arabs brought compass from China to west
AD 1150	China	Sternpost rudder invented
AD 1187	Alexander Neckam	First mention of a mariner's compass in *De Utensilibus*
AD 1290	Pisan Chart	Oldest extant surviving sea-chart (probably Genoese); first to show use of rhumb lines on a map
AD 1323	William of Ockham	Occam's Razor formulated in *Summa Logicae*
AD 1420	Henry the Navigator	Establishes Academy of Geography (Portugal)
AD 1421	Gavin Menzies	*The Year China Discovered the World* © 2003
AD 1430	Zheng He	Chinese fleet reaches east Africa
AD 1500	Leonardo da Vinci	Continuity in incompressible flow; "Remember, when dis-coursing on water, to induce first experience, then reason."
AD 1513	Juan Ponce de Leon	First description of Florida Current
	Piri Reis	Portolan chart of North and South Atlantic Ocean
AD 1534	Oronce Fine	Heart-shaped world map showing America and the Pacific
AD 1543	Nicolas Copernicus	Heliocentric theory
AD 1553	Gemma Frisias	Determining longitude with absolute time proposed
AD 1559	Haci Ahmet	Turkish cordioform projection map of world
AD 1567	William Borne	Common log invented; "knot" coined (1 nm/hour)
AD 1569	Gerardus Mercator	Map projection where compass bearings are a straight line
AD 1580	William Borough	First measured magnetic variation (England)
AD 1584	Lucas Waghenaer	Atlas portfolio; depths in fathoms reduced to half tide level
AD 1592	Elizabeth I	English Parliament established statute 1 mile = 5280 ft
AD 1600	William Gilbert	Dipole theory of Earth's magnetic field
AD 1609	Johann Kepler	Three laws of planetary motion
	Hugo Grotius	*Mare Liberum* published
AD 1621	Willebrord Snell	Established law of optical refraction
AD 1629	Rene Decartes	Developed analytic geometry, conservation of momentum
AD 1633	Edmund Gunter	Established that Earth's magnetic field is changing
AD 1638	Galeli Galileo	Principles of accelerated motion; founded mechanics
AD 1643	Evangelista Torricelli	Mercury barometer invented
AD 1654	Leopoldo Cardinal dei Medici	Cofounder of the *Accademia dei Cimento* - Florence; sealed liquid-in-glass thermometer invented
AD 1666	Robert Hooke	Freezing point of pure water basis of temperature scale
AD 1686	Edmond Halley	Primary winds and relation to ocean currents systematized
AD 1687	Isaac Newton	*Principia Mathematica* published; universal gravitation

Date	Person	Event
AD 1699	Isaac Newton	Earth proved an oblate spheroid; invented reflecting octant
AD 1701	Edmond Halley	Magnetic variations charted; isogonic line invented
AD 1720	Jacques Bellin	French Hydrographic Office established (first such office)
AD 1725	Luigi Marsigli	Monograph published including volumes on bottom topography, tides, currents, and temperature
AD 1735	George Hadley	Used conservation of momentum to explain Trade Winds
AD 1738	Daniel Bernoulli	Published *Hydrodynamica* on statics and fluid dynamics
AD 1740	Daniel Bernoulli	First study of equilibrium tides
	Leonard Euler	Most significant tide generating force tangent to Earth
AD 1742	Anders Celsius	Centigrade scale of temperature devised
AD 1747	Nils Gissler	Inverted barometer effect discovered
AD 1752	Philippe Buacher	Showed continuity between land and seafloor topography
AD 1755	Lisbon Earthquake	Destruction of Academy of Geography Library
	Leonard Euler	Fluid mechanics described in purely analytic terms
AD 1757	Thomas Godfrey	Marine sextant invented (first designed and named by Tycho Brahe in early seventeenth century)
AD 1760	John Harrison	Marine chronometer developed; £20,000 awarded in 1773
AD 1762	John Canton	Proved that water is slightly compressible
AD 1767	Nevil Maskelyne	*British Nautical Almanac* first published
AD 1772	Johann Lambert	Conformal conic projection for charting coastal areas
	James Cook	Circumnavigation of Antarctica at about 60°S
AD 1775	Pierre Laplace	Hydrodynamic equations including horizontal Coriolis terms
AD 1777	Benjamin Franklin	Gulf Stream map published
AD 1781	Joseph Lagrange	Introduced velocity potential and stream function in fluid motion
AD 1788	Charles Blagden	Discovered freezing point depression in sea water
	Joseph Lagrange	*Mecanique Analytique* published
AD 1802	Nathaniel Bowditch	*The New American Practical Navigator* published
	Franz Gerstner	First theory of surface waves in deep water
AD 1805	Adrien Legendre	Method of least squares published
AD 1806	Francis Beaufort	Devised 0-12 category marine wind scale
AD 1807	Thomas Jefferson	US Survey of the Coast established
AD 1814	Friedrich von Humbolt	Explained deep cold water in tropics as polar convection
AD 1818	John Ross	Self-registering thermometers for deep measurements
AD 1829	Robert Wauchope	Time ball invented, England
AD 1830	US Navy	Depot of Charts and Instruments established
AD 1831	William Redfield	Deduced rotary character of hurricanes and plotted paths
	Michael Faraday	Sea water motion through Earth's magnetic field may give measurable signals
AD 1832	James Rennel	Distinguishes between drift currents and stream currents
AD 1835	Gaspard de Coriolis	Fluid motion on a rotating Earth

(Continued)

(*Continued*)

Date	Person	Event
AD 1839	James Clark Ross	Deep sea sounding of 2677 fathoms off Africa
AD 1842	Georges Aimé	Invented the reversing thermometer
	Julius Mayer	Theory of mechanical nature of heat and equivalent to work one year before James Joule (1818–1889)
AD 1843	Christian Doppler	Velocity of moving source affects frequency perceived by observer
AD 1845	George B. Airy	Theory of infinitesimal surface gravity waves
AD 1846	George Stokes	"Stokes Theorem" fluid circulation = integral of vorticity
AD 1847	Hermann von Hemholtz	Conservation of energy conceptualized
AD 1850	George Stokes	"Stokes Law" on settling velocity of spheres in a fluid
AD 1851	Leon Foucault	Proved the rotation of Earth by pendulum experiment
AD 1853	James Coffin	Observed the relationship between wind and pressure which became known as "Buys-Ballot's Law", after chief of Dutch Meteorological Services Christoph Buys-Ballot (1854–1889)
AD 1855	Matthew Maury	*Physical Geography of the Sea* published; sailing directions published (cf. Coast Pilots of US Coast and Geodetic Survey)
AD 1865	Johann Forchhammer	Concentration ratios of sea water constituent ions constant
AD 1872	*HMS Challenger*	Began four year voyage obtaining many deep-sea soundings; quantum jump in oceanic knowledge
	William Thomson	(Lord Kelvin) mechanical tide prediction machine
	J.B. Listing	Coined term geoid to describe Earth's equipotential surface (a surface of equal potential energy)
AD 1874	James Croll	Showed wind stress and heating needed for ocean currents
AD 1876	George Stokes	Phase velocity of deep water waves twice the group velocity
	C.M. Guldberg and H. Mohn	Concept of linear friction introduced: $\alpha F = Jv$
AD 1878	William Thomson	Sounding machine perfected using a sintered glass tube and self-acting sounder, piano wire, and winch
	Rutherford B. Hayes	US Coast and Geodetic Survey established
AD 1879	William Thomson	Tidal harmonic analyzer
	Josef Stefan	Total radiant energy proportional to T^4 (T in kelvin)
AD 1881	Samuel Langley	Developed bolometer for measuring solar irradiance
AD 1883	Osborne Reynolds	Laminar and turbulent flow transition—the Reynolds Number
AD 1885	Henrik Mohn	Derived geostrophic formula
AD 1890	John Elliot Pillsbury	Observations of Florida Current velocity structure from *USS Blake*
AD 1893	Fridtjof Nansen	Began three-year polar drift in RV Fram; Nansen Bottle developed (*ca.* 1925 according to other authorities)

Date	Person	Event
AD 1896	Valentine Boussinesq	Introduced concept of eddy coefficients of viscosity
AD 1898	Vilhelm Bjerknes	Introduced method of "dynamic sections" to geostrophy
AD 1901	Martin Knudsen	Hydrographical Tables for salinity and density
AD 1902	Vagn Walfrid Ekman	Earth's rotation deflects currents 45° cum sole to wind
AD 1904	US Navy	First broadcast of time signals to correct chronometers
	Rollin Harris	Tidal amphidromes proposed
AD 1905	Erik Fredholm	Time-dependent Ekman currents
AD 1911	Alexander Behm	Used sound from an explosion to measure water depth
AD 1912	Bjørn Helland-Hansen	Deep in situ temperatures increase by compressional heating
	SS Titanic	International Ice Patrol established
AD 1916	Bjørn Helland-Hansen	Introduced the temperature-salinity (T-S) diagram
AD 1919	Sir Napier Shaw	Introduced the adjective "geostrophic" in *Manual of Meteorology*, Volume 6, © Cambridge University Press
AD 1922	US Navy	First practical echo sounder
	Louis Richardson	Numerical integration of equations of motion for forecasting
AD 1924	Georg Wüst	Geostrophic and direct currents in Straits of Florida
AD 1925	*RV Meteor*	Used acoustics extensively to chart deep-sea for two years
	RV Discovery	Commissioned for Southern Ocean research
AD 1929	William Morrison	Used vibrations of quartz crystals to measure time
AD 1933	Jonas Fjedstad	Theory of internal waves
AD 1938	Athelstan Spilhaus	Mechanical bathythermograph (BT) invented
	Milutin Milankovitch	Earth's climate controlled by orbital variations
	Irving Langmuir	Surface motion of water induced by wind
AD 1939	Carl-Gustaf Rossby	Conservation of potential vorticity applied to long waves
AD 1940	Columbus Iselin	Gulf stream volume transport variability
AD 1942	Sverdrup, Johnson, and Flerming	*The Oceans, Their Physics, Chemistry, and General Biology* first published
	MIT	LORAN invented
AD 1947	Harald Sverdrup	Theory of wind-driven currents in a baroclinic ocean
AD 1948	Henry Stommel	Explained westward intensification of ocean currents
AD 1954	Townsend Cromwell	(with Montgomery and Stroup) Equatorial undercurrent
AD 1957	IGY	International Geophysical Year
AD 1960	Bissett-Berman Co.	Conductivity, Temperature, Depth (CTD) invented
	United Nations Educational, Scientific and Cultural Organization (UNESCO)	Intergovernmental Oceanographic Commission established

(Continued)

(Continued)

Date	Person	Event
AD 1965	USC&GS/ Intergovernmental Oceanographic Commission (IOC)	Tsunami Warning System in the Pacific established
	US DoC	Environmental Science Services Administration established
AD 1966	Lady Bird Johnson	USC&GS Ship *Oceanographer* christened
AD 1970	US DoC	National Oceanic and Atmospheric Administration (NOAA) established
AD 1978	NASA	SEASAT launched; first satellite altimeter
	NASA	NIMBUS-7 launched; first colorimeter (CZCS)
AD 1981	NOAA-n	Advanced Very High Resolution Radiometer launched
AD 1985	US Navy	GEOSAT launched; collinear orbits invented
AD 1990	International	World Ocean Circulation Experiment (WOCE) started
AD 1992	NASA	TOPEX/Poseidon launched (sea surface height)
AD 1993	US DoD	Global Positioning System (GPS) fully operational
AD 1997	NASA	NASA Scatterometer (NSCAT) launched (surface winds)
AD 1997	NASA	SeaWiFS Launched (ocean color)
AD 1997	NASA	Tropical Rainfall Measuring Mission (TRMM) launched
AD 1998	US Navy	GEOSAT Follow-On (GFO) Mission launched
AD 1998	United Nations	International Year of the Ocean
AD 2000	Greenwich Observatory	Last year of the twentieth century

Additional Reading

Brown, L.A. 1979. *The Story of Maps.* New York, NY: Dover Publications, 397 pp.

Bowditch, N. 1966. *American Practical Navigator.* Washington, DC: U.S. Navy Hydrographic Office, H.O. Pub. No. 9, 1524 pp.

Cotter, C.H. 1968. *A History of Nautical Astronomy.* New York, NY: American Elsevier, 387 pp.

Davis, N. 1979. *Voyages to the New World.* New York, NY: Wm. Morrow & Co, 287 pp.

Deacon, G.E.R. 1962. *Seas, Maps, and Men.* New York, NY: Doubleday, 297 pp.

Eco, U. and G.B. Zorzoli. 1963. *The Picture History of Inventions.* New York, NY: MacMillan, pp. 123–132.

Freuchen, P. and D. Loth. 1957. *Peter Freuchen's Book of the Seven Seas.* New York, NY: Julian Messner, Inc, 512 pp.

Heyerdahl, T. 1979. *Early Man and the Ocean: A Search for the Beginning of Navigation and Seaborne Civilizations.* New York, NY: Doubleday & Co, 438 pp.

Map: Exploring the World, 2015. New York. London: Phaidon Press, Inc., 352 pp.

Marx, R.L. and J.G. Marx. 1992. *In Quest of the Great White Gods.* New York, NY: Crown Publishers, 343 pp.

Maul, G.A. 1998. A Brief history of timekeeping. In: P. J. Fell, M. Kumar, G. A. Maul, and G. Seeber (editors), *Fourth International Symposium on Marine Positioning, INSMAP* 1998. ©Marine Technology Society, Washington DC, pp. 459–470.

Menzies, G. 2003. *1421–The Year China Discovered the World.* New York, NY: Harper Collins, 576 pp.

Skelton, R.A., T.E. Marston, and G.D. Painter. 1965. *The Vinland Map and the Tartar Relation.* New Haven, CT: Yale University Press, 291 pp.

Sobel, D. 1995. *Longitude.* New York, NY: Walker & Co, p. 184 pp.

Stommel, H. 1965. *The Gulf Stream*, 2nd edition. Berkeley, CA: University of California Press, 248 pp.

Trychare, T. 1973. *The Lore of Ships.* New York, NY: Crescent Books, 279 pp.

Van Sertima, I. 1976. *They Came before Columbus.* New York, NY: Random House, 288 pp.

von Arx, W.S. 1962. *An Introduction to Physical Oceanography.* Boston, MA: Addison-Wesley Publishing Co, pp. 351–395.

Wede, K. 1972. *The Ship's Bell, Its History and Romance.* New York: South Street Seaport Museum, 16 Fulton Street, 62 pp.

Whitfiled, P. 1996. *The Charting of the Oceans.* Rohneret Park, CA: Pomegranate Art books, 136 pp.

Appendix B: Glossary of Important Terms for Seagoing Scientists and Engineers

AAUS: the American Academy of Underwater Sciences; organization promoting dive training and safety for seagoing scientist and engineers.

ABCD: airway, breathing, circulation, defibrillation; acronym for rendering CPR.

Abeam: relative direction 090° from dead ahead (000°) or 270° from dead ahead; measured clockwise.

ABS: American Bureau of Shipping; a vessel classification society for insurance and safety.

ADCP: Acoustic Doppler Current Profiler; measures vertical profile of horizontal water currents beneath the ship's hull.

AED: Automatic External Defibrillator; device for administrating CPR if there is no pulse.

AGOR: Auxiliary General Purpose Oceanographic Research; US Navy research vessels.

AIS: Automatic identification system; computerized ship-to-ship transceiver communicates call letters, course, speed, etc.

Altitude: vertical angle at the observer between the horizon and a celestial body.

AM: amplitude modulation in telecommunications; ante meridian (before mid-day) in time keeping.

Amidships: area on ship near that above the keel at any deck level; "rudder amidships" is a command to the helmsman to align the rudder with the keel (0° rudder angle).

Amphidrome: point on ocean-basin-scale tidal wave where there is no amplitude.

Amplitude: horizontal angle of the sun at sunrise or sunset (celestial navigation); half of wave height (physical oceanography).

AP: Assumed Position; latitude and longitude used in celestial navigation to simplify arithmetic.

ARGOS: joint France/US surface location and tracking system from satellites.

Autopilot: electro-mechanical helm for setting in a course and self-steering.

AUV: Autonomous Underwater Vehicle; a programmable unmanned submersible.

Aweigh: referring to the anchor being raised clear of the seafloor; "anchors aweigh."

Awl: sailmaker's handheld tool with wooden handle and sail-needle eye at the tip.

Azimuth: horizontal angle of an object with respect to true north (true azimuth), magnetic north (magnetic azimuth), or the bow (relative azimuth).

Azimuth Circle: flanged sighting ring that fits on a compass bowl for taking a bearing.

Barycenter: mass center of Earth–Moon system; forms smooth elliptical solar orbit.

Baseline: intersection of MLW with the land; datum of the EEZ.

Bathymetry: the art and science of measuring seafloor topography.

Bathyscaphe: extreme deep-diving manned research vessel.

Bathythermograph: instrument for measuring temperature versus depth while making way.

Beaufort Scale: wind speed and sea state (wave height) scale used by marine observers.

Bearing: horizontal angle from a ship to an object; may be true, magnetic, or relative.

Bilge: area between the hull and "A" deck or tank tops.

Bilge Keel: a fore-and-aft turn-of-the-bilge external nonstructural fin to reduce rolling.

Binnacle: housing for a magnetic compass or a gyrocompass.

Bits: two short vertical deck posts placed two post-diameters apart for securing mooring lines.

Bitter End: end of a line requiring whipping (natural fiber) or shrink wrap (synthetic fiber).

Block: mechanical apparatus for smoothly changing the direction of line or wire rope.

Boat: a vessel whose length overall (LOA) is 65 feet (20 meters) or less, so defined as to adhere to the US Motorboat Act of 1940.

Boatswain: senior noncommissioned or nonlicensed deck seaman; pronounced "bos'n."

Bow: forward-most structure of a boat or ship.

Bowline: arguably the most useful sailor's knot; pronounced "boh-lin."

Bridge: pilothouse or wheelhouse; compartment where the ship is steered.

Bucket Station: a small platform hinged at deck level and lowered to support the oceanographer while handling equipment on a "hydro" wire.

Bulkhead: vertical plates separating compartments (ships do not have walls).

Bulwark: vertical main-deck waist-high safety bulkhead fitted with freeing ports and chocks.

Bunk: sleeping furniture (ships do not have beds).

Buoy: unmanned floating surface vessel used for navigation or information gathering.

Buoyancy: ability of a vessel to float; a force in vessel stability (Archimedes, *ca.* 287-211 BC).

Ca: *circa*; approximate date.

Cabin: stateroom, office, head, and sometimes galley of the CO.

Capstan: vertical rotating, flanged, powered drum for handling line, especially mooring lines.

Cardinal: north, east, south, and west compass directions.

Chafing: wearing of line or wire rope due to rough surfaces in a chock or fairlead.

Chart: projection of marine or air navigation information on a flat surface.

Chief Scientist: person in charge of the scientific party and ship's scientific program.

Chock: an opening in a bulwark, usually built to prevent chafing.

Chronometer: a marine instrument for timekeeping that has a determinable rate of change.

Cleat: horns-shaped deck fitting for securing a line in a figure-eight pattern.

CME: Chief Marine Engineer; USCG-licensed officer in charge of ship's machinery and engineroom personnel.

CO: Commanding Officer (captain of a commissioned ship or facility).

Coaming: a raised frame to hold a hatch cover, and strengthen the deck opening.

Coasting: visual navigation along a coast.

Coast Pilot: NOAA publication with information detailing port entrance and harbor facilities.

Collision Bulkhead: forward-most vertical athwartships structure isolating bow from first compartment, and watertight from the keel to the main deck.

COLREGS: international regulations for preventing collisions at sea.

Commissioning: a naval ceremony formally naming a ship, where the sponsor often breaks a bottle of champagne on the bow.

Compartment: enclosed interior structure such as a stateroom.

Compass: nautical instrument for indicating direction; may be gyro or magnetic.

CPO: Chief Petty Officer; highly rated enlisted person on a commissioned ship.

CPR: cardiopulmonary resuscitation; method in first aid for restoring heartbeat and breathing.

CQM: Chief Quartermaster; senior enlisted navigation specialist on a commissioned ship.

CTD: conductivity, temperature, depth (pressure actually); instrument for obtaining a vertical profile of the ocean, suspended from an electrical conductor cable using a slip-ring winch.

Datum: horizontal or vertical reference surface on which a chart is projected.

Dead Reckoning: position estimation by projecting course and speed of vessel.

Deck: horizontal structural surface with several levels; "A-deck" is that just above the bilge.

Declination: angle at Earth's center north or south of the equator of a celestial object.

Depth: vertical distance from the main deck to the keel (naval architecture); vertical distance from the sea surface to the seafloor (charting).

Deviation: horizontal angle between magnetic north and compass north, measured east or west.

Displacement: weight of water displaced by volume of a ship; expressed in long tons.

Dive Ladder: ladder leading from the dive platform into the sea; usually aft of the transom.

Dogs: handles that pivot to secure a weather-tight door.

Draft: vertical distance from a ship's keel to the waterline.

Ebb Current: a tidal current flowing from the land to the ocean.

ECDIS: Electronic Chart Display and Information System; a computer system IMO-approved to replace paper charts.

EEZ: Exclusive Economic Zone; 200 nm distance from baseline surrounding a country over which the state has jurisdiction including scientific research.

EPIRB: Emergency Position Indicating Radio Beacon; automatically sends distress signal.

ESSA: Environmental Science Services Administration (1965–1970); predecessor to NOAA.

Eulerian: mathematical perspective observing seawater or air from a fixed point; formulated by Swiss mathematician Leonhard Euler (1707–1783).

Fairlead: device for smoothly changing the direction of a line or wire rope; may have a sheave.

Fantail: open deck area near the stern from which many ocean science and ocean engineering activities are conducted.

Fathom: six feet; to understand as in "can you fathom that?"

Fathometer: electro-acoustic display system for measuring and displaying water depth.

First Point of Aires: celestial longitude of the vernal equinox.

Fix: geographic position determination: visual, celestial, electronic.

Fleet Angle: angle at winch toward the block or fairlead needed to ensure smooth spooling of cable on the drum.

Flood Current: tidal current flowing from the sea toward the land.

Floor: vertical beam in the bilge between the hull and tank top.

Flying Bridge: deck immediately above the bridge or pilothouse.

Freeboard: vertical distance between the main deck and the waterline.

Freeing Port: opening in a bulwark at deck level allowing seawater runoff.

Forecastle: forward-most area of open deck at the bow; pronounced "fo'c'sle."

FRS: Fellow of the Royal Society, a singular honor in the United Kingdom.

Funnel: vertical engine exhaust structure, typically that of a steamship.

Galley: compartment for preparing food.

GHA: Greenwich hour angle; angle along equator between Greenwich meridian and a celestial body's meridian; measured westward from Greenwich.

Gimbals: two concentric rings hinged so as to keep an instrument level as the ship moves.

Give-way Vessel: ship that must change course or speed under COLREGS.

GM: metacentric height; critical transverse stability criterion to prevent capsizing.

GMT: Greenwich Mean Time; time at the Prime Meridian; replaced by UTC.

GOES: Geostationary Operational Environmental Satellite; operated by NOAA.

GP: geographic position (of a celestial body).

GPS: Global Positioning System; range-range location by satellite trilateration.

Gravimeter: gravity meter for measuring Earth's total gravitational acceleration.

Great Circle: line on a sphere's surface formed by the intersection of a plane through the sphere's center.

Gudgeon: ring-like structure for accepting the pintle of a rudder.

Gypsyhead: horizontal-rotating flanged drum for handling line usually on the windlass.

Gyro: slang for gyrocompass, electro-mechanical device for indicating true north.

HA: hour angle; angle along the equator between two celestial longitudes; measured westward.

Halyard: a line used to raise flags, sails, and/or a spar.

Hatch: vertical opening through a deck for personnel or cargo.

Hauling Part: section of a block-and-tackle line used to raise or lower a load.

Hawse Pipe: slanted tube at bow to house anchor shank.

Hawser: large diameter heavy-duty line used in mooring a large vessel.

Head: shipboard sanitary compartment with sink and commode.

Heave: vertical vessel motion.

Helm: device, usually a wheel, for steering a ship.

Helmsman: person steering a ship.

HMS: His (or Her) Majesty's Ship; government vessel of the United Kingdom.

Horizon: a great circle always 90° from the observer's zenith.

Hull: exterior structure enclosing all spaces at least up to the main deck.

ICOADS: International Combined Ocean-Atmosphere Data Set; a rich resource of environmental sea surface data from NCAR.

IMO: International Maritime Organization, a United Nations agency focused on SOLAS.

Intercardinal: northeast, southeast, southwest, and northwest compass directions.

IOC: Intergovernmental Oceanographic Commission of UNESCO; a United Nations agency dedicated to international marine science and calibration standards (*cf.* WMO).

IRSO: International Research Ship Operators, a voluntary organization of marine research vessel-operating institutions.

JOOD: Junior Officer of the Deck; OOD in training.

Keel: centerline fore-and-aft primary vertical structural member of a ship.

Knot: in speed, one nautical mile per hour; in seamanship, a method of joining or fastening a line (*cf.* bowline).

Ladder: structure with rungs or steps for vertical assent or descent (ships do not have stairs).

Lagrangian: mathematical perspective of following a parcel of sea water or air; formulated by Italian mathematician Joseph-Louis Lagrange (1736–1813).

LAN: local apparent noon; exact time when sun is highest in the sky.

Latitude: angular distance at Earth's center 0°-90° north or south of the equator.

Lazarette: storage compartment under the fantail just forward of the transom.

Leadline: a sounding instrument consisting of a lead weight and a line marked in feet or fathoms; "mark twain" equals two fathoms water depth using a leadline.

LHA: local hour angle; angle along equator between the local meridian and a celestial body's meridian; measured westward from observer.

Line: in seamanship a linear flexible piece of natural or synthetic fiber; "crossing the line" refers to going across the equator on a ship (where there may be a ceremony converting a pollywog to a shellback).

List: angle from vertical of an upright ship.

LMT: local mean time; exact time at observer's meridian.

LOA: length overall of a ship, measured from the forward-most point of the bow (stem) to the aft-most point of the stern.

Locker: storage furniture for equipment and/or personal articles.

Log: record of ship operations and oceanographic station data; "to log it" is to make a written record; the "ship's log" is a legal document.

Longitude: angular distance along the equator 0°–180° east or west of the Greenwich (England) meridian.

Lookout: seaman qualified to watch and listen for danger under COLREGS.

LOP: line of position; locus of all possible positions in a given direction.

Lubber's Line: vertical mark on a marine compass, parallel to the keel, to allow reading of a compass card and steering a course.

MARAD: US Maritime Administration, an agency of the US Department of Transportation.

Marlinspike: tapered tool of wood or metal used in working with line.

Masthead Light: white navigation light showing from dead ahead to 2 points abaft the beam on both sides (20 point light).

Mate: USCG-licensed deck watch officer.

Meridian: great circle on Earth's surface through the geocenter and the poles.

Meridian Angle: angle along equator between the local meridian and a celestial body's meridian; measured eastward or westward from observer.

Mess: compartment onboard where meals are served.

Messenger: hand-sized weight clipped to the hydro wire for tripping Nansen Bottles.

Metacenter: a point to which the center of gravity may rise and the ship will still possess positive transverse stability.

Minute: in time, one sixtieth of a mean solar hour; in angle (arc-minute) one sixtieth of a degree.

MHW: mean high water; tidal datum on US nautical charts for elevations and clearances.

MLLW: mean lower low water; tidal datum on US nautical charts for water depth.

MOB: man overboard.

Monkey Fist: weighted ball-shaped knot on bitter end of a heaving line.

Moon-pool: a vertical shaft through the hull to the main deck for access to the sea.

Mousing: wrapping wire around the shackle bow and pin to prevent the pin from loosening.

MSI: marine safety information; products of NGA.

MSV: mean sounding velocity; geometric mean used in bathymetry needed for correcting acoustic signals for sound speed variations in the water column.

Muster: gathering of personnel for emergencies or meetings, and where attendance is taken.

Nadir: point directly below observer (*cf.* zenith).

Nansen Bottle: a metal sampling instrument about 2 feet tall fitted with reversing thermometers, valves, petcocks, and a messenger release mechanism.

Napier's Diagram: plot of ship's magnetic deviation.

NCAR: the National Center for Atmospheric Research.

NGA: National Geospatial-Intelligence Agency (US Department of Defense).

Niskin Bottle: a plastic sampling instrument using bungee to close the end-caps rather than valves; capacity ranges from 1–50 L.

nm: nautical mile (6076 ft or 1852 m); equal to 1 arc-minute of latitude.

NOAA: National Oceanic and Atmospheric Administration, an agency of the US Department of Commerce.

NOS: National Ocean Service, a division of NOAA.

NWS: National Weather Service, a division of NOAA.

Offing: area offshore of a geographical feature.

Offshore: seaward of the coast; in meteorology a wind from the land toward the sea.

One-Hand: a cry for assistance at sea.

Onshore: landward of the coast; in meteorology: a wind from the sea toward the land.

OOD: Officer of the Deck (senior watch-stander on the bridge of a commissioned ship).

Overhead: horizontal structure atop bulkheads often hiding piping and wiring (modern ships do not have ceilings).

Period: time taken for a ship to roll from port to starboard and return; time from crest to crest of a progressive wave.

Perpendicular: imaginary vertical line where the bow intersects the water-line and the stern (usually the rudder post) to which the "length between perpendiculars" is referenced.

PFD: personal floatation device.

Pilot: highly expert local master mariner who guides ships in and out of a port or harbor.

Pilot Chart: chart showing monthly mean wind, wave, barometric pressure, surface temperature, and so on for an ocean basin.

Pinger: acoustic device designed to send two signals simultaneously, one toward the bottom, and one toward the surface ship.

Pintle: vertical pin for suspending the rudder; sits into the gudgeon.

Pitch: rotational vessel motion fore and aft; also angle of propeller blade.

PLB: personal locator beacon; cellphone-sized emergency radio transmitter.

Plimsoll Mark: load line marking welded onto the hull amidships on commercial vessels to indicate the maximum safe draft for expected weather and sea conditions.

PM: post meridian (after mid-day) in time keeping.

PMO: Port Meteorological Officer; a professional employee of NWS.

Point: compass direction in multiples of 11¼ degrees.

Pollywog: a seaman who has not crossed the equator ceremoniously.

Porthole: circular opening through the hull or superstructure with glass insert.

Portside: left-hand side of ship facing forward.

Projection: geometry of drawing a sphere (Earth) onto a flat surface (a chart).

Propeller: rotating screw-like structure for fore-and-aft thrust.

RA: right ascension; angle along the equator 0°–360° east of Greenwich meridian.

RADAR: radio detection and ranging; electromagnetic transmit/receive system for detecting objects through fog or mist.

RATS: Research Application Tracking System; US Department of State system for monitoring applications to conduct oceanographic research in foreign waters.

RDF: radio direction finder; electronic instrument for obtaining bearings on a radio transmitter.

Rode: components of anchoring system including anchor chain, shackle, and anchor line.

Roll: vessel motion about the longitudinal axis side to side.

Rope: twisting of strands of natural fiber in opposite direction of threads to make finished product; synthetic fibers may be woven.

ROV: remotely operated vehicle; a submersible tethered to a control room with cabling.

Rudder: external pivoting vertical structure near stern for directing ship to port or starboard.

RV: research vessel of discovery operating on the sea.

SAFE: stop, assess, find, exposure; action acronym before rendering first aid.

SAMPLE: signs, allergies, medications, past, last, events: acronym of information to be gathered by the first responder and transmitted to the medical professionals.

SAR: search and rescue (spoken as "S," "A," "R").

Scale: ratio of distance on a chart to distance on Earth; 1:100,000 is cited as "1 to 100,000" for example.

SCUBA: Self-Contained Underwater Breathing Apparatus.

Sea: locally generated wind waves (*cf.* "swell"); often with whitecaps and foam.

Seabag: container for carrying personal items including toiletries and clothing.

Sea Chest: volume between the hull and tank top through which sea water is pumped; often fitted with valves; may be instrumented.

Seaman: in admiralty law, any person working on a ship; male or female "deck hand."

Seamanship: essential nautical skills for seagoing scientists and engineers (and others).

Second: in time, one sixtieth of a minute; in angle (arc-second), one sixtieth of an arc-minute.

Sextant: navigation instrument for measuring angles by an observer; accurate to 0.1 arc-minute.

SHA: sidereal hour angle; celestial longitude of a body with respect to the First Point of Aires.

Sheave: wheel on a block for line or wire to feed across (ships do not have pulleys).

Shellback: a seaman who has crossed the equator ceremoniously.

Ship: a vessel whose LOA is over 65 feet, defined so as to adhere to the US Motorboat Act of 1940.

Shoreline: intersection of MHW with the land on a nautical chart.

Sidelight: navigation light; green to starboard; red to port; 10 point light showing from dead ahead to two points abaft the beam.

Sidereal: planetary and satellite motion with respect to the stars.

SNAME: Society of Naval Architects and Martine Engineers, an organization of professionals in ship design and operations.

SOLAS: safety of life at sea; international program for preventing harm or loss of life.

Sole: deck of the cockpit of a yacht.

SONAR: sound navigation and ranging.

Splice: weaving natural fiber to form an eye, an end, or joining two together (short splice, or long splice if the line's diameter must be kept constant); eye splice in synthetic fiber.

SSB: single side band; highly efficient amplitude modulation (AM) radio transmission method; 2182 kHz SSB is an international hailing and emergency frequency.

Stack: vertical housing surrounding the engine exhaust funnel(s).

Standing Part: section of block-and-tackle line between one block and another.

Stand-on Vessel: ship that must maintain course and speed under COLREGS.

Starboard: right-hand side of a ship facing forward.

Stateroom: sleeping quarters for the crew and scientific party (ships do not have rooms).

Stem: forward-most point of bow structure; idiom: "stem to stern."

Stern: after-most structure of a boat or ship; opposite to the bow.

Stern Light: white navigation light showing 12 points from dead astern to two points abaft the beam on either side.

Stern-vest: an inflatable minimal lifejacket used when working on an open deck.

Strand: twisting of yarns in opposite direction in making rope.

Stringer: horizontal fore-and-aft beams strengthening hull above the tank top.

Stuffing-box: horizontal tube through which the propeller shaft is fitted that prevents water from entering the ship.

Stuffing-tube: watertight horizontal tube through which electrical wire is passed.

Superstructure: decks and compartments on or above the main deck.

Surge: vessel vertical motion keel-to-superstructure.

Sway: vessel horizontal motion side-to-side.

Swell: long period waves from a distant source; usually no whitecaps.

Syzygy: when the Earth, moon, and sun all align (new moon/full moon).

Thimble: oval-shaped insert to an eye splice to spread the load evenly.

Thruster: athwartships tunnel with an impeller for creating side-to-side thrust.

Tonnage: space available for commercial purposes expressed in long tons (2240 pounds).

Transom: athwartships exterior bulkhead on stern structure just aft of the lazarette.

UNCLOS: United Nations Conference on the Law Of the Sea.

UNESCO: United Nations Educational, Scientific and Cultural Organization.

UNOLS: University-National Oceanographic Laboratory System, an organization of the US National Science Foundation, for managing the research fleet.

US Board on Geographic Names: agency of the US Department of Interior, chartered to provide official names; NOAA nautical charts strictly adhere to Board names.

USC&GS: United States Coast and Geodetic Survey, a former agency of the US Department of Commerce now separated into the National Ocean Service and the National Geodetic Survey of NOAA.

USCG: United States Coast Guard, an agency of the US Department of Homeland Security.

USMS: United States Maritime Service, an agency of the US Department of Transportation.

USNS: United States Naval Ship, a nonwarship of the US Navy.

USPHS: United States Public Health Service, an agency of the US Department of Health and Human Services.

USS: United States Ship, a warship of the US Navy.

UTC: coordinated universal time (French *temps universel coordonné*); international time standard replaces GMT.

Variation: horizontal angle between true north and magnetic north, measured east or west.

VCG: vertical center of gravity; point in transverse stability through which weights act.

Vessel: a vehicle capable of transportation on water; includes seaplanes.

VHF: Very High Frequency radio band (30 MHz to 300 MHz).

VOS: voluntary observing ship; vessel making surface weather reports every six hours GMT.

Wardroom: commissioned officers' mess presided over by the XO.

Watch: on-duty-time aboard ship.

Waterline: horizontal line on a ship where the water intersects the hull.

Whipping: wrapping sail twine around the bitter end of a line to prevent unravelling.

Winch: deck machine powered by steam, or electricity, or hydraulics with a drum for spooling wire, either hydrographic wire or electrical conductor cable.

Windlass: specialized winch for handling the anchor chain or rode.

WMO: World Meteorological Organization, a United Nations agency dedicated to international weather forecasting and global climate change quantification (*cf.* IOC).

XO: Executive Officer (second in command of a commissioned ship or facility).

Yarn: first twisting of natural fibers to make line; yarns are twisted into strands.

Yaw: vessel rotational motion port to starboard.
Zenith: point directly overhead in the navigational triangle.
Zone Time: hour in multiples of 15° longitude zones; abbreviated "ZT."

Additional Reading

Baker, B.B., Jr., W.R. Deeble, and R.D. Geisenderfer. 1966. *Glossary of Oceanographic Terms*, 2nd Edition. Washington, DC: Special Publication 35, US Naval Oceanographic Office, 204 pp.

Appendix C: Answers to Exercises for Seagoing Scientists and Engineers

Chapter 1

1. There are many websites with information about the voyage of the *HMS Challenger* (for example). One with good information about the ship itself is at: http://www.19thcenturyscience.org/HMSC/HMSC-INDEX/Deck-Plans.html.

2. The first "oceanographic institute" arguably is Station Biologique de Roscoff founded in France as a marine biological station in 1859. http://marinebio.org/oceans/history-of-marine-biology/

3. What is it? Brass signal cannon cast in Spain in 1702 and lost in the wreck of the 1715 treasure fleet off central east Florida.

Chapter 2

1. Good leadership is *leadership by example*! Introduce yourself, and go around the room having everybody introduce themselves including their job during the voyage. Explain the purpose of the cruise and the value of the work in which they are engaged. Have a prepared *watch list* to pass out and ask if changes are needed; accommodate changes if possible being sure that the skills necessary for each watch are spread as needed. Post the revised watch list on the scientific bulletin board. Set a strict *no drugs or alcoholic beverages* policy, and that each person is expected to be 10 minutes early in relieving the watch, properly dressed for duty. Explain the "line-of-command" but encourage friendly interaction with the ship's crew—they have much to offer! Open the meeting to questions and discussion. Never ask a member of the science party to do something that you yourself will not do—leadership by example.

2. Online cardiopulmonary resuscitation (CPR) courses are available, but hands-on training is quite essential. All seagoing scientists and engineers should be certified every two years. See http://www.redcross.org/take-a-class/cpr for courses and dates.

3. What is it? Sounding is the process of determining water depth; this *sounding tube* has a metal frame and a sintered glass pipe open at one end for water to compress the trapped air. A nonlinear gauge compliments the unit.

Chapter 3

1. The University-National Oceanographic Laboratory System (UNOLS) website lists research vessels of four class sizes plus certain National Oceanic and Atmospheric Administration (NOAA) and US Coast Guard (USCG) ships. Much of the language of steel ship construction is readable from deck plans at: https://www.unols.org/.

2. Etymology is the study of word origins and how they have changed over time. The phrase "stem to stern" contains two such words, but the technical meaning of "stem" has changed with the transition from wooden ships to steel ships. The *Oxford Dictionary of Word Origins* is an amazing resource for your research.

3. What is it? Starboard sidelight from a nineteenth century sailing ship, originally lighted by a lantern fueled by whale oil.

Chapter 4

1. The variables for calculating the change in metacentric height (GM) are

	Weights (tons)	VCG (feet)	Moments
Research Vessel (Δ)	100	8	800
Laboratory Van (W)	3	13	39
Difference (d)	103	5	839

The change in G to G′ is $\dfrac{W \times d}{\Delta} = \dfrac{3 \times 5}{100} = 0.2$ ft, and the new G′M = 4.0 − 0.2 = 3.8 ft. Thus, the vessel has adequate transverse stability for the added load of the van, but the roll period $T = \dfrac{0.44 \times \text{beam}}{\sqrt{GM}}$ would be slightly longer depending on the beam (*b*).

2. "Oh, hear us when we cry to Thee. For those in peril on the sea!"— William Whiting (1860); see ftp://ftp.soest.hawaii.edu/dkarl/misc/dave/UH&theSea/R-Chapter14.pdf as an example.

3. What is it? Draftsman's set for drawing in pencil and for inking with "India Ink"; materials are plated brass and ivory.

Chapter 5

1. The Global Positioning System (GPS), developed by the US Department of Defense in the late 1970s. A constellation of 24 satellites in 20,000 km altitude orbits broadcast precise position and time signals that are received on Earth and in low orbit satellites such as TOPEX/Poseidon. Simultaneous reception of signals from four or more satellites allows solution of four equations in four unknowns: $(x - x_i)^2 + (y - y_i)^2 + (z - z_i)^2 = ([t_i - b - s_i]c)^2$, $i = 1, 2, 3, 4$ ... where x_i y_i z_i, and s_i are the position and time from four (or more) satellites, c is the speed of light, and x y z and $t_i - b$ are the position and clock-corrected time of the receiver respectively. The term $(t_i - b - s_i)c$ is the distance from the satellite to the receiver. Einsteinian relativistic effects are accounted for in order for GPS to function properly; numerous US Government and commercial Internet sites add much detail; see www.gps.gov for example, and be sure to research differential GPS (dGPS).

2. Refer to Figure 3.4 and imagine a pier running parallel to the portside. Bits are shown as a pair of circular structures—two on the bow (port and starboard), one on each side amidships, and two on the stern – port quarter and starboard quarter. Using your skills of relative bearings, the bowline is toward 315° from the port bow bits toward the pier; the stern line is toward the port quarter bits toward 225°, and the breast line is from 270° from the amidships bits. The after bow spring leads from the bow bits aft at an angle of about 200°, and the forward quarter spring from the stern bits toward 340° or so. With the wind forcing the ship against fenders between the ship and the pier, the captain will want to "spring" the stern out by going forward against the after bow spring: full left rudder, slow ahead on the starboard propeller and once the stern is away from the pier, slow astern on the port propeller. Once the stern is about 45° from the direction of the pier, the engines will be stopped, the after bow spring slipped, and the ship will back away from the pier. Draw it out! Research "dipping the eye" as a means of allowing two ships to share the same bollard on the pier.

3. Dr. Roswell Austin of the Scripps Visibility Lab invented an ocean color scale to replace the Forel Scale and the Ule Scale of the nineteenth century. His scale used standard color palettes in a plastic case, and was used in early studies of ocean color from space from instruments such as the CZCS—the coastal zone color scanner—which flew on the Nimbus 7 spacecraft.

Chapter 6

1. The magnetic variation is changing 9 arc-minutes per year, so from 1988 to 2016 the variation is $(2016 - 1988)$ years $\times 9\dfrac{'}{year} = 252' \div 60\dfrac{'}{°} = 4°12' + 3°45' = 7°57'$ or for practical purposes the variation now (2016) is 8°W. The table from Chapter 6 is

GPS Course	Variation	Magnetic Course	Deviation	Compass Course
352	8 W	000	10 W	010
022	8 W	030	15 W	045
052	8 W	060	20 W	080
082	8 W	090	25 W	115
112	8 W	120	10 W	130
142	8 W	150	0	150
172	8 W	180	5 E	175
202	8 W	210	10 E	200
232	8 W	240	15 E	225
262	8 W	270	5 E	265
292	8 W	300	5 W	305
322	8 W	330	7 W	337
352	8 W	360	10 W	010

2. True course 090° to compass: $090°T + 8°W = 098°M + 21°W = 119°C$. The 21°W comes from a linear interpolation between 25°W at 090° magnetic and 10°W at 120° magnetic. Next true course 180° to compass: $180°T + 8°W = 188°M - 6°E = 182°C$. Recognizing this as a 3:4:5 triangle, the relative course back to the sea buoy is $\theta = \arctan\dfrac{30}{40} = 37°$ and the true course is $360° - 37° = 323°T$. The compass course is $323°T + 8°W = 331°M + 7°W = 338°C$. Draw the triangle with the courses to complete the assignment.

3. "What is it?" is an Ekman (Vagn Walfrid Ekman, 1874–1954) inspired mechanical current meter. It used a magnetic compass to record direction by dropping BB-sized lead balls into a hopper, where the number of balls was proportional to the current speed in that direction. The compass had to be corrected from magnetic to true.

Chapter 7

1. To plan this voyage, or any other voyage, start with the nautical chart of the area. Go to http://www.nauticalcharts.noaa.gov/mcd/ ccatalogs.htm and select the largest scale chart that covers the area (remember the larger the scale the smaller the area). The chart will give information needed to decide where to locate oceanographic stations. Be sure to study the chart legend to determine the depth units (feet or fathoms for US charts). Remember to use the latitude scale to measure distances, never the longitude scale. Time = distance ÷ speed (nautical miles ÷ knots). Remember that the Gulf Stream between Florida and the Bahamas averages 3 kn; calculate the crab angle $\theta = \arctan \dfrac{3 \text{ kn}}{7 \text{ kn}}$, and the distance across the Straits of Florida to be run as $d = \dfrac{3}{\sin\theta}$ for each nautical mile between Miami and Bimini.

2. For a small area such as 1° latitude by 1° longitude, Earth can be considered a plane. The north-south distance $\Delta y = 111.1 \dfrac{\text{km}}{\circ} = 111.1 \times 10^3$ m and the east-west distance $\Delta x = 111.1 \dfrac{\text{km}}{\circ} \times \cos 35.5 = 90.4 \times 10^3$ m. Thanks to Pythagoras (*ca.* 570–495 BC), we can calculate the distance from $\phi = 36°N, \lambda = 75°W$ to $\phi = 35°N, \lambda = 74°W$ as $\sqrt{(90.4)^2 + (111.1)^2} = 143.2$ km. Thus, the pressure gradient force per unit mass from $\phi = 36°N, \lambda = 75°W$ to $\phi = 35°N, \lambda = 74°W$ is $9.8 \text{ m} \times \text{s}^{-2} \times \dfrac{1 \text{ m}}{143.2 \times 10^3 \text{ m}} = 6.8 \times 10^{-5} \text{ m} \times \text{s}^{-2}$ (for the physical oceanographer, the geostrophic current $c_g = \dfrac{6.8 \times 10^{-5} \text{m} \times \text{s}^{-2}}{1,4584 \times 10^{-4} \sin 35.5\text{s}^{-1}} = 0.8 \dfrac{\text{m}}{\text{s}} \times 1.94 \dfrac{\text{kn}}{\text{m} \times \text{s}^{-1}} = 1.6$ kn, toward the northeast, a reasonable Gulf Stream speed).

3. The objects shown as "What is it?" are spacing dividers, sometimes called 11-point dividers. They are used to evenly space distances between two points, and are stainless steel in construction.

Chapter 8

1. From top to bottom: Masthead light is white and shines from right ahead to two points (22½°) abaft the beam on either side. Port sidelight is red and shines from right ahead to two points abaft the port beam. Starboard sidelight is green and shines from right ahead to two points abaft the starboard beam. Stern light is white and shines from dead astern to two points abaft the beam on both sides. In addition, this vessel will have one all-round white light when at anchor (and a black ball in daytime), and all-round "restricted in ability to maneuver" lights, red over white over red (or during daytime—black ball over black diamond over black ball). A vessel when seeing ahead a red sidelight will know they are the give-way vessel; similarly when seeing a green sidelight they assume that they are the stand-on vessel. When seeing only the white stern light, that they are the overtaking vessel, and must keep out of the way.

2. The small boat operator must have a device for attracting attention in case of emergency. During daytime, this includes a horn or a flag or a gun or a whistle; at night it can include a flare, a horn, or a flashing light. Using Channel 16 for an immediate emergency—such as a fire out of control or a flooding—the signal spoken is "mayday, mayday, mayday." If the situation is dangerous but not of imminent threat to life, the signal spoken is "pan, pan, pan." Historically the signal is SOS (established in the early 1900s as the method of communicating distress in Morse Code; replaced in 1999 by the Global Maritime Distress and Safety System), or in Morse code: dot, dot, dot, dash, dash, dash, dot, dot dot, which would be sent by radiotelegraph. Be sure to have your position and ship's name when making such a distress message.

3. The tooth-rattling "What is it?" is a nineteenth century fog horn. The horn is used in foggy conditions during restricted visibility, day or night, to announce the existence of your ship.

Chapter 9

1. Working Load Limit (WLL) is defined by the International Rope Access Trade Association (IRATA.org) as "the maximum mass or force which a product is authorized to support in general service when the pull is applied in-line, unless noted otherwise, with respect to the centerline of the product; that is, the WLL of a component is specified by the manufacturer." In practice the WLL is the minimum breaking load MBL divided by a safety factor SF (usually 5 for wire rope and 10 for fiber): $WWL = MBL \div SF$. Note in the definition, the

WLL is in-line use, so if used in a sling, the value will change. Also note that the definition applies to new rope. The table below is for 3/8 inch line (www.bevisrope.com):

Material	WWL (Pounds)
Nylon twisted rope	287
Polyester/Dacron (Poly/Dac) composite rope	240
Polyester twisted rope	334
Polypropylene twisted rope	340
Manila rope	122
Sisal, cotton rope	108
Double braid nylon rope	980

A nice example of all the factors needed in designing safe lifting of a load including the eye bolts, hooks, chain, and line is given in http://ehs.whoi.edu/ehs/occsafety/LoadCalcs.pdf.

2. Knots and splices for a variety of applications can be learned at (for example) http://www.animatedknots.com/. Try to learn them with your eyes closed. Can you make a one-handed bowline (it might be useful to tie it around your chest in case you are a man overboard victim and want to be lifted back onboard)?

3. The "What is it?" pictured is a pelican hook on a nylon strap, As with line, rope, chain, eyes, and so on, hooks have WLLs as well. It is the manufacturer's responsibility to provide test-based data and WLL recommendations.

Chapter 10

1. Identify the parts: frame, tongue, hitch coupler, axle, wheels, tires, fenders, guide posts, jack, winch, bunks, hub; missing: spare tire, tire nut wrench, stands, chains, blocks ...

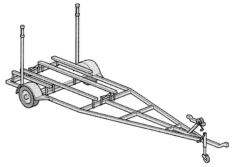

Before hitching, check to see that the coupler and the ball are matched. After aligning the ball and coupler, lower the tongue with the trailer jack until the ball is seated; secure the ball in place with a pin or lock. Cross the chains from the coupler to the hitch placing the eye of the hook toward the trailer. Connect the wiring harness and check to see that both the brake lights and the turn signal lights work. Check to see that the boat is secured with two tie-down straps—one forward of amidships, and one aft. Remove any loose items that may fly out of the boat while on the road. Most trailer hubs will have a zerk fitting for adding fresh grease to the bearing assembly; add grease until it oozes out.

Before changing a tire, be sure that the rig is fully off the road and that there is ample safety room for the work. Place safety reflectors on the roadside. Block the opposite tire. Loosen the lug nuts with the lug-nut wrench, jack up the frame, place jack stands, remove the lug nuts and tire, place the new wheel on the studs, replace the lug nuts so that the chamfer is inward, hand-tighten the lug nuts, remove the jack stands and lower the trailer, tighten the lug nuts in a star pattern, shake the wheel to be sure it is seated, remove the blocks, check the lights and chains, and continue on the trip. Repair the tire as soon as possible.

2. Rear Admiral John Elliot Pillsbury (1846–1919) was a US Navy lieutenant when he commanded the Coast and Geodetic Survey steamer *George S. Blake* (one of only two US ships to have their names inscribed in the façade of the *Musée Océanographique*, Monaco—can you name the other?). Much of his work can be found in the rare books collection of NOAA (the US Coast and Geodetic Survey is a NOAA parent agency). One source of Pillsbury's innovative work is in http://docs.lib.noaa.gov/rescue/oceanheritage/Gc296g9p541891. pdf. Following the introduction of tapered steel cable for deep-sea dredging by Alexander Agassiz (1835–1910), Pillsbury anchored the *Blake* in depths over 2000 fathoms using wire cable, chain, and anchor. The cable is sections of wire rope from 3/8 inch diameter to 7/16" to ½" to 5/8" and the anchor is of the "Cape Ann" style weighing about 500 pounds with a stock of hardwood.

3. "What is it?" is a wire angle indicator often used on the hydro wire, but useful for anchoring as well.

Chapter 11

1. The maximum fleet angle θ is 2°, the deck bolts are UNOLS standard 2 foot spacing. Scale the distance of the A-frame aft of the transom to be 6 feet, the deck to be 18 feet, the winch set 2 feet father forward;

$6 + 18 + 2 = 26$ feet from the winch to the meter wheel on the A-frame. The distance to the meter wheel must be 15 times the width (w) of the drum: accordingly $\tan \theta = \dfrac{\text{opposite}}{\text{adjacent}} = \dfrac{w/2}{26} = \tan 2°; w = 1.8$ ft, or for practical purposes a maximum 2' width drum. Check the drum width in Figure 3.4.

2. For modern steel wire rope typical data are as follows:

Diameter (inches)	Maximum Strength (lb$_f$)	Working Load (lb$_f$)
¼	5480	1100
½	21400	4280
¾	47600	9520
1	83600	16700

Maximum safe mass for a 3/8″ (9.5 mm) wire rope, where the safe load is 2440 lb$_f$ $\left(2440\ lb_f \times 4.45\dfrac{N}{lb_f} = 10{,}900N\right)$ is

$$\text{mass} = \frac{\text{force}}{\text{gravity}} = \frac{10.9 \times 10^3\,N}{9.8\ m-s^{-2}} = 1112\ kg \times \frac{2.2\ lb}{kg} = 2446\ lbs \div \frac{2000\ lb}{\text{short ton}}$$

$$= 1.2\ \text{tons.}$$

3. The *Fram* Expedition was organized and led by Fridtjof Nansen, for whom the "What is it?" is named: the Nansen Bottle. The left and right hand thermometers are protected reversing thermometers, and the center one is an unprotected (from water pressure) thermometer.

Chapter 12

1. NOAA nautical chart 11469 is at a scale of 1:100,000, and nicely covers the crossing. Soundings are in fathoms and feet; convert 1 fathom $= 6$ ft $\div \dfrac{3.28\ \text{ft}}{m} = 1.83$ m. Plot the stations (using 11-point dividers (see answers to Chapter 7) if available), scale off the ϕ and λ, and build a table of station number, ϕ, λ, water depth, and estimated time $(\text{minutes}) = 2 \times \dfrac{\text{depth}}{50\dfrac{m}{\text{minute}}}$ to take each cast for the captain and scientific party. If rosette samples are to be taken, add the additional time for that procedure, say one extra minute per Niskin bottle.

2. *Observing Handbook No. 1* is available from http://www.vos.noaa.gov/ObsHB-508/ObservingHandbook1_2010_508_compliant.pdf. Training as a PPT is available at http://www.vos.noaa.gov/training.shtml. The instruments that you will need include: sling psychrometer, bucket and line, sea surface thermometer, anemometer, compass, and a stopwatch. The Beaufort Scale is preferred as a means of estimating wind speed and sea state, but teach the maneuvering board solution as well

3. What is it? Before electric calculators and digital computers, station calculations were made using a slide rule. This particular example is of the Culbertson Circular Slide Rule used to calculate thermometric depth from corrected reversing thermometers (see the Nansen Bottle "What is it?" from Chapter 11).

Chapter 13

1. Project instructions must be very specific and complete. The captain of the research vessel must be able to plot the stations on a chart or enter the station coordinates into the electronic chart display and information system (ECDIS). Since this voyage includes a foreign country, you as chief scientist must contact the US Department of State at least 6 months prior to the planned cruise and initiate RATS—the Research Application Tracking System (http://go.usa.gov/3mAMm). RATS information should be included in the project instructions.

2. Tides and tidal currents are provided by the NOAA Center for Operational Oceanographic Products and Services (http://tidesandcurrents.noaa.gov). Searching for "Gun Cay" will not return information, but searching for "Caribbean Islands" will open a long list of sites with information. For Gun Cay, consult the nautical chart for nearby islands and choose the closest (North Cat Cay in this case). The NOAA site will provide up to one month of predictions. If farther into the future or into the past, the publication *Tide Tables East Coast of North and South America including Greenland* can be purchased or found in larger libraries.

3. Modern tide gauges are electronic instruments, but "What is it?" is a standard (mechanical) US tide gauge *ca.* 1900. It would be housed in a tide house (about 8′×8′ square), have a stilling well and float and counterbalance weights, and receive daily visits from the tidal observer. *Instructions for Tide Observations* by G.D. Rude, US Coast and Geodetic Survey Special Publication 139, might be of interest for history buffs.

Chapter 14

1. Your write-up should address the following issues and others that you deem necessary for the safety and success of the trip: equipment to load before getting started, towing vehicle choice (4-wheel drive preferred), hook-up to boat and trailer, choosing the launching site, choosing the land route to the launching site, launching procedure and parking of tow vehicle/trailer, water route from launching marina to the flood shoal, navigation, anchoring while the divers are in the water, anchor ball and dive flag "alpha" (see Figure 3.2), safety diver, recovering the anchor and route back to the marina, reloading the boat onto the trailer, returning to the boat storage area, cleanup and storage of equipment. Most universities will have a form from the Dive Safety Officer; be sure it is filled out well in advance and posted.

2. http://www.boatus.org/courses/ is a source of a free boating course that (for example) satisfies the requirements of the Florida Fish and Wildlife Conservation Commission (http://myfwc.com/boating/) for Florida boaters. Many commercial sites are also available with costs ranging up to $30. In-class instruction is also available; the interested should check with the local Power Squadron or the Coast Guard Auxiliary.

3. The classic Clarke-Bumpus plankton sampler is Chapter 14's "What is it?" It allowed quantitative estimates of plankton populations at specific depths using a messenger to operate the instrument.

Chapter 15

1. Assume our Phoenician navigator lifts his inclinometer (the astrolabe wasn't invented yet) at noon when the sun is highest above the northern horizon and records the solar attitude as 31½°. The navigator knows that this day is the summer solstice so the declination of the sun is 23½°N. The Phoenician's latitude is $\phi = 90° - 23.5° - 31.5° = 35°S$. If it were the winter solstice ($\delta = 23.5°N$), the sun's altitude would have been $h_0 = 90° - 35° + 23.5° = 78.5°$ above the northern horizon.

2. $16^h07.9^m$ EDT $= 20^h07.9^m$ GMT, and from the *Nautical Almanac* (http://navsoft.com/downloads.html) for April 15, 2000, using linear interpolation, $\delta = 10°05.8'N$ and $GHA\odot = 122°02.2'$. The assumed longitude $\lambda = 80°W$, and so the local hour angle and meridian angle respectively are $LHA\odot = 122°02.2' - 80° = 42°02.2'$ and $t = 122°02.2' - 80° = 42°02.2'W$. Solving $\cos(co - h_c) = \cos(co - \delta)$

$\cos(co - \phi) + \sin(co - \delta)\sin(co - \phi)\cos(t)$ for the computed altitude, $\sin h_c = \sin\delta\sin\phi + \cos\delta\cos\phi\cos t$, or $h_c = 46°42.7'$. The altitude intercept is $h_o - h_c = 46°48.9' - 46°42.7' = 6.2' = 6.2$ nm Using computed greater away (CGA), the LOP is 6.2 nm toward the GP of the sun, whose azimuth from $\dfrac{\sin(Z)}{\sin(co - \delta)} = \dfrac{\sin(t)}{\sin(co - h_c)}$ is N106°W = 254°T. That is from the assumed position of 28°N/80°W, the LOP is

6.2 nm in the direction $Z_n = 254°$, and is orthogonal to the azimuth Z_n. Use Figure 15.2 to plot your sun line geometry.

3. The globe in the "What is it?" box is a spherical star chart whereby the navigator sets the Greenwich Hour Angle (GHA) of Aires and his/her latitude to display stars above the horizon.

Index